SpringerBriefs in Public Health

SpringerBriefs in Public Health present concise summaries of cutting-edge research and practical applications from across the entire field of public health, with contributions from medicine, bioethics, health economics, public policy, biostatistics, and sociology.

The focus of the series is to highlight current topics in public health of interest to a global audience, including health care policy; social determinants of health; health issues in developing countries; new research methods; chronic and infectious disease epidemics; and innovative health interventions.

Featuring compact volumes of 55 to 125 pages, the series covers a range of content from professional to academic. Possible volumes in the series may consist of timely reports of state-of-the art analytical techniques, reports from the field, snapshots of hot and/or emerging topics, literature reviews, and in-depth case studies. Both solicited and unsolicited manuscripts are considered for publication in this series.

Briefs are published as part of Springer's eBook collection, with millions of users worldwide. In addition, Briefs are available for individual print and electronic purchase.

Briefs are characterized by fast, global electronic dissemination, standard publishing contracts, easy-to-use manuscript preparation and formatting guidelines, and expedited production schedules. We aim for publication 8–12 weeks after acceptance.

Monica Eriksson • Lenneke Vaandrager
Bengt Lindström
Editors

The Hitchhiker's Guide to Salutogenesis

From the Ottawa Charter for Health Promotion to Planetary Health

Second Edition 2025

Center of Salutogenesis

Editors
Monica Eriksson
Department of Psychology
Lund University
Lund, Sweden

Lenneke Vaandrager
Department of Social Sciences
Health & Society
Wageningen University & Research
Wageningen, The Netherlands

Bengt Lindström
Norwegian University of Science and
Technology (NTNU)
Trondheim, Norway

ISSN 2192-3698   ISSN 2192-3701 (electronic)
SpringerBriefs in Public Health
ISBN 978-3-031-89567-8   ISBN 978-3-031-89568-5 (eBook)
https://doi.org/10.1007/978-3-031-89568-5

1st edition: © Bengt Lindström and Monica Eriksson 2010

This work was supported by The Center of Salutogenesis (COS) at the University of Zürich.

© The Editor(s) (if applicable) and The Author(s) 2025. This book is an open access publication.

Open Access This book is licensed under the terms of the Creative Commons Attribution-NonCommercial-NoDerivatives 4.0 International License (http://creativecommons.org/licenses/by-nc-nd/4.0/), which permits any noncommercial use, sharing, distribution and reproduction in any medium or format, as long as you give appropriate credit to the original author(s) and the source, provide a link to the Creative Commons license and indicate if you modified the licensed material. You do not have permission under this license to share adapted material derived from this book or parts of it.
The images or other third party material in this book are included in the book's Creative Commons license, unless indicated otherwise in a credit line to the material. If material is not included in the book's Creative Commons license and your intended use is not permitted by statutory regulation or exceeds the permitted use, you will need to obtain permission directly from the copyright holder.
This work is subject to copyright. All commercial rights are reserved by the author(s), whether the whole or part of the material is concerned, specifically the rights of reprinting, reuse of illustrations, recitation, broadcasting, reproduction on microfilms or in any other physical way, and transmission or information storage and retrieval, electronic adaptation, computer software, or by similar or dissimilar methodology now known or hereafter developed. Regarding these commercial rights a non-exclusive license has been granted to the publisher.
The use of general descriptive names, registered names, trademarks, service marks, etc. in this publication does not imply, even in the absence of a specific statement, that such names are exempt from the relevant protective laws and regulations and therefore free for general use.
The publisher, the authors and the editors are safe to assume that the advice and information in this book are believed to be true and accurate at the date of publication. Neither the publisher nor the authors or the editors give a warranty, expressed or implied, with respect to the material contained herein or for any errors or omissions that may have been made. The publisher remains neutral with regard to jurisdictional claims in published maps and institutional affiliations.

This Springer imprint is published by the registered company Springer Nature Switzerland AG
The registered company address is: Gewerbestrasse 11, 6330 Cham, Switzerland

If disposing of this product, please recycle the paper.

Preface

Aaron Antonovsky was the first to introduce a theory and model that generates health. This was named salutogenesis and explains the origin of health. It was based on a health continuum spanning from the health end of the ease/dis-ease continuum. It was later connected to health promotion. Today there are other theories following salutogenic pathways; therefore, we can talk about a salutogenic umbrella including models and theories using a salutogenic approach to health and quality of life.

The first edition of *The Hitchhiker's Guide to Salutogenesis* was published in 2010, whereafter the research area has been widely accepted in most health sciences and accepted as a theory foundation for Health Promotion. Since then, the number of scientific publications has expanded almost exponentially. The book has been translated into nine languages. The first edition was based on the two books by Aaron Antonovsky, *Health, Stress, and Coping* [1] and *Unraveling the Mystery of Health. How People Manage Stress and Stay Well* [2], and an extensive systematic research synthesis, 1992–2003, *Unravelling the Mystery of Salutogenesis. The evidence base of the salutogenic research as measured by Antonovsky's Sense of Coherence Scale* [3] and several publications up to 2010.

This second edition of *The Hitchhiker's Guide to Salutogenesis* is revised and expanded with new areas of research, is peer reviewed by international scientists in the field, as well as updated until 2024. It is written to explain salutogenesis mainly in terms of Antonovsky's theoretical model because this is where it started and where we still have the best knowledge and evidence today. Salutogenesis is also one of the strongest theories for the promotion of health. Rather than going much into detail there is an attempt try to keep the text simple, explaining the core, giving cross reference to books and scientific articles on the existing evidence. Further, the book explains salutogenesis through images and metaphors, some of them originating from Aaron Antonovsky himself.

Thereafter much has happened. Two editions of *The Handbook of Salutogenesis* [4, 5] present the global evidence on salutogenesis. These handbooks are open access publications which, together with the previously mentioned literature, are the foundation for this second edition of *The Hitchhiker's Guide to Salutogenesis*.

Further, there are two additional books especially valuable for explaining salutogenesis, that is, *Salutogenic organizations and change, The concepts behind organizational health intervention research* [6], and *Health Promotion in Health Care - Vital theories and research* [7].

The latter is a unique, short yet comprehensive publication explaining the salutogenic theory and further including key examples of its implementation, mainly serving as an introduction not only for newcomers but also for teachers and students of any profession attached to health and social sciences.

It is important to understand that salutogenesis is a broader concept than merely the measurement of sense of coherence. One thing should be clear already from the beginning: there are no simple shortcuts. You cannot just take a ride and hitchhike straight to salutogenesis. The implementation of salutogenic principles demands a change of mind and a thorough integration of thoughts and action. It takes time to set new coordinates and use a new compass in reorienting mindsets, but this could be one of the most meaningful things to improve health promotion performance. We do think what is in this book can give good guidance and once you get the knack of it, there is no turning back.

Have good journey!

Gothenburg, Sweden Monica Eriksson
Wageningen, The Netherlands Lenneke Vaandrager
Bornholm, Denmark Bengt Lindström
December 2024

References

1. Antonovsky, A. (1979). *Health, stress and coping*. Jossey-Bass.
2. Antonovsky, A. (1987). *Unraveling the mystery of health. How people manage stress and stay well*. Jossey-Bass.
3. Eriksson, M. (2007). *Unravelling the mystery of salutogenesis. The evidence base of the salutogenic research as measured by Antonovsky's sense of coherence scale*. Folkhälsan Research Centre.
4. Mittelmark, M. B., Bauer, G. F., Vaandrager, L. et al., (Eds.). (2017). *The handbook of salutogenesis*. Springer.
5. Mittelmark, M. B., Bauer, G. F., Vaandrager, L. et al., (Eds.). (2022). *The handbook of salutogenesis* (2nd edn). Springer.
6. Bauer, G. F., & Jenny, G. J. (2013). *Salutogenic organizations and change. The concepts behind organizational health intervention research*. Springer.
7. Haugan, G. & Eriksson, M. (Eds.). (2021). *Health promotion in health care – Vital theories and research*. Springer.

Acknowledgements

To all the *co-authors*, thank you for your enthusiasm and the time you spent on the content of the book, despite your heavy workload in your daily commitments.

To the *international team of peer reviewers*, your feedback was essential for us to improve the proposal to Springer. Thank you for all the comments and the time you spent on this publication.

To our *colleagues around the world*, especially to the Editor-in-Chief of the two *Handbooks of Salutogenesis, Professor Maurice Mittelmark*, thank you for your encouragement and support to write a second edition of *The Hitchhiker's Guide to Salutogenesis*.

To the *production team at Springer* and the *Executive Editor Janet Kim*, we are grateful for your contribution, the advice, and a never-ending patience to guide us in every stage of the production process.

To Peter Wikström, Stil & Form, for realizing our sketches into professional graphic design (Figures 1.1–1.3, 2.2–2.4).

On behalf of readers worldwide, the editors acknowledge with gratitude the open-access financial support by the *Center of Salutogenesis at the University of Zürich*. This creates opportunities for anyone interested to read the book and will help to further disseminate salutogenesis in practice.

On behalf of the Editors

Monica Eriksson
Lenneke Vaandrager
Bengt Lindström

Contents

Part I Background

1 **Salutogenesis: A Compass for Health Promotion** 3
   Bengt Lindström, Monica Eriksson, and Lenneke Vaandrager

2 **A Change of Perspective** 13
   Bengt Lindström

3 **The Original Salutogenic Framework** 21
   Monica Eriksson

4 **The Orientation to Life Questionnaire: Sense of Coherence** 35
   Monica Eriksson

5 **Health, Mental Health and Quality of Life** 47
   Monica Eriksson and Eva Langeland

Part II Applications

6 **Salutogenesis in the Context of Health Care** 59
   Monica Eriksson and Eva Langeland

7 **Salutogenesis in the Context of Learning Processes** 69
   Lenneke Vaandrager, Maria Koelen, and Laura Bouwman

8 **Salutogenesis in the Context of Work** 79
   Georg F. Bauer and Anja I. Lehmann

9 **Salutogenesis in the Context of Society** 89
   Ruca Maass, Lenneke Vaandrager, and Jake Sallaway-Costello

10 **From the Ottawa Charter to Planetary Health** 103
   Jake Sallaway-Costello, Claudia Meier Magistretti, and
   Bengt Lindström

Part III An Outlook

11 **Critical Issues Related to the Salutogenic Theory and Its Implementation** 117
Laura Bouwman and Lenneke Vaandrager

12 **Future Perspectives** 127
Bengt Lindström, Monica Eriksson, Lenneke Vaandrager, and Georg F. Bauer

Appendix: Resources and Meeting Places of Salutogenesis: The Global Working Group, the Society, the Handbook and the Center of Salutogenesis. 133

Index 137

About the Editors and Contributors

Editors

Monica Eriksson, PhD is a Professor in public health and health promotion, and, from 2025, an affiliated researcher at Lund University, Sweden. She was educated at Åbo Akademi University in Vasa, Finland, where she is also an Associate Professor in social policy. She was a member of the IUHPE Global Working Group on Salutogenesis (2007–2018). Her research focus, from the beginning, has been on salutogenesis. In 2007, she defended her doctoral thesis, a systematic research synthesis, based on more than 450 scientific papers on studies using Aaron Antonovsky's Sense of Coherence scale. Her research on health resources on different levels and on various samples continues to further elaborate the understanding of salutogenesis.

Lenneke Vaandrager, PhD in health promotion, is an associate professor in Health and Society at Wageningen University and Research, The Netherlands. She takes on a sociological perspective to study the healthy living environment. Her work—in research, managerial and teaching roles—is underpinned by the values of care, participation, and justice. She works together with landscape architects, social geographers, plant scientists, public health nutritionists, health promotion professionals, policy makers, and citizens. Her overall research focus is to analyze and contribute to the development of inclusive healthy settings: contexts in which people engage in daily activities and in which environmental, organizational, and personal factors interact to affect health and wellbeing. She loves working around the theme of nature and wellbeing: green citizen initiatives, livable cities, and green care. She is a member of the Global Working Group on Salutogenesis and a coordinator of the European Training Consortium in Public Health and Health Promotion (ETC-PHHP).

Bengt Lindström, MD, PhD, DrPH, Professor of Salutogenesis, Public Health and Health Promotion (retired 2017), first trained as a Specialist in Pediatrics. The original interest in a broader understanding of health came from his research on quality of life in children developing the wellbeing dimension of health. He worked with Aaron Antonovsky in Antonovsky's last seven years of life and initiated research and academic training in salutogenesis immediately after Antonovsky's premature death in 1994, thereby continuing the new tradition of regarding health as a resource and constructive concept known for innovative and graphic ways of explaining health. He is affiliated with several universities in Northern Europe and initiated and ran the IUHPE Global Working Group on Salutogenesis until his retirement in 2017.

Contributors

Georg F. Bauer Center of Salutogenesis, Division of Public and Organizational Health, Epidemiology, Biostatistics and Prevention Institute, University of Zürich, Zürich, Switzerland

Laura Bouwman Health & Society, Wageningen University & Research, Wageningen, The Netherlands

Monica Eriksson Department of Psychology, Lund University, Lund, Sweden

Maria Koelen Health & Society, Wageningen University & Research, Wageningen, The Netherlands

Eva Langeland Department of Health and Caring Sciences, Faculty of Health and Social Sciences, Western Norway University of Applied Sciences, Bergen, Norway

Anja I. Lehmann Center of Salutogenesis, Division of Public and Organizational Health, Epidemiology, Biostatistics and Prevention Institute, University of Zürich, Zürich, Switzerland

Bengt Lindström Norwegian University of Science and Technology (NTNU), Trondheim, Norway

Nordic School of Public Health, Gothenburg, Sweden

Ruca Maass Department of Neuromedicine and Movement Science, Norwegian University of Science and Technology (NTNU), Trondheim, Norway

Claudia Meier Magistretti Institute for Early Childhood Education, University of Graz, Graz, Austria

Jake Sallaway-Costello Division of Food, Nutrition & Dietetics, University of Nottingham, Nottingham, UK

Lenneke Vaandrager Health & Society, Wageningen University & Research, Wageningen, The Netherlands

List of Abbreviations

ASPHER	Association of Schools of Public Health in the European Region
CAM	Complementary and Alternative Medicine
CFA	Confirmatory Factor Analysis
CSOC	Childrens' Sense of Coherence
DS	Depressive Symptoms
ETC	European Training Consortium
ETC-PHHP	European Training Consortium in Public Health and Health Promotion
EUHPID	European Health Promotion Indicator Development
FSOC	Family Sense of Coherence
GRR	Generalized Resistance Resource
HFA	Health For All
HHG	The Hitchhiker's Guide
HP	Health Promotion
HQoL	Health-related Quality of Life
HR	Human Rights
IPA	Interpretative Phenomenological Analysis
IUHPE	International Union for Health Promotion and Education
JD-R	Job Demands-Resources Health model
MDG	Millennium Development Goals
OC	Ottawa Charter
OLQ	Orientation to Life Questionnaire
PAR	Participatory Action Research
QoL	Quality of Life
RCT	Randomized Controlled Trial
SAL	Salutogenesis
SalWork-N	Salutogenic Survey on Sustainable Working Life for Nurses
SDG	Sustainable Development Goals
SFC	Sense For Coherence
SHAPE	Salutogenic Healthy Aging Program Embracement

SHIS	Salutogenic Health Indicator Scale
SMH	Salutogenic Model of Health
SOC	Sense of Coherence
SOCC	Sense of Community Coherence
SRR	Specific Resistance Resource
STARS	Society for Theory and Research on Salutogenesis
SWPS-SF	Salutogenic Wellness Promotion Scale, short form
WEMS	Work Experience Measurement Scale
WHO	World Health Organization
Work-SoC	Work-related Sense of Coherence

Part I
Background

Chapter 1
Salutogenesis: A Compass for Health Promotion

Bengt Lindström, Monica Eriksson, and Lenneke Vaandrager

In 2010 a short book on salutogenesis was published by Monica Eriksson and Bengt Lindström named the *Hitchhiker's Guide to Salutogenesis: Salutogenic Pathways to Health Promotion* [1]. This book was a condensed version of the PhD thesis of Monica Eriksson, titled *Unravelling the Mystery of Salutogenesis*. This was the first systematic review that was published on salutogenesis [2]. While collecting the material and analysing what had been published on salutogenesis between 1992 and 2003 it turned out there were many researchers and professionals that published in the name of salutogenesis not knowing or following the original framework nor did they use the original research tools. This was seen as a risk to the potential and development of the research area. Aaron Antonovsky, who unfortunately suffered an unexpected early death in 1994, had stated the importance of not changing the method of measuring the sense of coherence before there was enough evidence to show it stands on a sound and solid evidence base. This led to the need for the systematic review of the whole research area.

Simultaneously, the Global Public Health agenda was undergoing a fundamental change of shifting from activities mainly devoted to classic public health supporting the healthcare system to also involving communities and professions outside the healthcare sector in a participatory community action in the promotion of health.

B. Lindström (✉)
Norwegian University of Science and Technology (NTNU), Trondheim, Norway

Nordic School of Public Health, Gothenburg, Sweden

M. Eriksson
Department of Psychology, Lund University, Lund, Sweden
e-mail: monica.eriksson@psy.lu.se

L. Vaandrager
Health and Society, Wageningen University and Research, Wageningen, The Netherlands
e-mail: lenneke.vaandrager@wur.nl

Further, the focus was on generating better health conditions thus improving the wellbeing and quality of life of the population and making the community a better place to live. One problem for health promotion was the lack of any theoretical foundation. Pairing the two new innovations and explaining how they could create synergy and gain from each other was a central mission of the first hitchhiker's guide. Further, there was a need to analyse the evidence of the five Action Areas of Health Promotion with a salutogenic framework to prove its effectiveness related to the Ottawa Charter. It seemed quite evident that health promotion could benefit if framed with a salutogenic perspective. WHO also wanted to develop education in the field of health promotion and commissioned ASPHER in 1990 to develop teaching programmes for the Action Areas of the Ottawa Charter. The Action Area "Lifestyle" was completed and in the continuation, this later became a still ongoing training programme based on health promotion and salutogenesis named The European Training Consortium in Public Health and Health Promotion (ETC). The third editor of the revised Hitchhiker's Guide (HHG), Lenneke Vaandrager, was a participant in this first course in 1991 and Aaron Antonovsky was teaching in the second one.

Today health promotion and salutogenesis have developed new practices and research. However, the world has also gone through an enormous change and become a less stable, fragile reality where human rights often are violated. Famish and wars still are a plague to mankind causing unwanted mass migration with fatal consequences. The human influence has become a profound challenge for the life on earth ultimately threatening our health and total existence. A coordinated effort by the global community is to build a sustainable response and address issues of climate change and move beyond only the human perspective to planetary health. This book aims to reopen and rewrite the script of the HHG and create a new salutogenic compass for our future development.

Health Promotion of 1986

The central document of health promotion, the Ottawa Charter (OC), was constituted at an international WHO health conference in Ottawa in 1986 [3]. To understand the OC one has to understand the historical context of modern public health and health promotion dating back to the time right after the Second World War and the foundation of the United Nations and the Declaration of Human Rights. WHO was in a sense instituted to defend Human Rights from the perspective of health.

Between 1948 and 1977 WHO moved from a health care and health system approach towards a contextual population approach to health. This meant a huge change of perspective initiated and put into effect by the visionary WHO Director Halfdan Mahler. What was to come could already be sensed in the push for primary health and primary health care at the WHO/UNICEF Alma Ata Conference in 1977 [4]. However the first contours of health promotion were seen in the WHO Global Strategy Health for All by the year 2000 [5]. The vision was to reduce inequity, form a sustainable development, use an interdisciplinary and intersectoral approach and aim not only at "adding years to life" but also "life to years." The latter is the first notion of wellbeing and quality of life becoming the outcome of a new direction in public health.

A coherent document of the principles of health promotion was prepared by WHO in 1984 [3]. The Ottawa Charter includes a set of principles and values, outcome formulations and strategic plans and a set of five action areas. The development after Ottawa is presented in a series of WHO reports [6]. In the Charter, health promotion is defined as

> the process enabling people to gain control over their health determinants thereby improving their health in order to be able to lead an active and productive life.

and visually presented in Fig. 1.1.

The human being in the centre represents mankind. In terms of the Declaration of Human Rights (HR) and the Ottawa Charters main principle, empowerment, we function as *active participating subjects in our own life*. The spiral represents the lifelong learning process of health. To the left, the health determinants (HD) are visualized. These are a given at birth (such as genetics, socioeconomic conditions, family and social capital, culture and traditions …). Throughout life, we constantly face new life events and gain life experiences (LEs) that shape the consistency of our lives and the resources for our ability to manage life. In the picture, the dotted line is pointing at a phenomenon (could be a health problem) we need to solve. To solve it we draw on our resources in the backpack containing our individual resources. These can also be found in our setting or context where we can draw on our external resources. We either solve or fail to solve the problem. Both experiences can become health resources for the future when we encounter a similar phenomenon. In terms of the Ottawa Charter, the main objective of the overall process of health development is to enable us to lead an active and productive life (APL). This could as well be called the Good Life or Quality of Life. Thus, quality of life

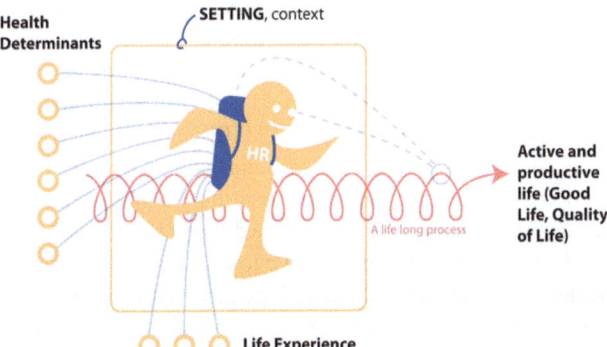

Fig. 1.1 The core of the Ottawa Charter—the individual is an active and participating subject where health is a lifelong process [1]. (Reprinted with permission of © Bengt Lindström. All Rights Reserved)

can be seen as the implicit outcome of the health process. At the heart, health was seen as a process enabling people to develop health through the control of health determinants thereby giving people the opportunity to lead an active and productive life. The community approach and policies leading to a healthy society became central thus expanding the focus from individuals and groups to the context of life. There was however no clear theoretical framework supporting the Charter. This later caused problems for the health promotion movement.

If this reasoning is turned into salutogenic terms, the process is about *comprehending* what health resources (or health determinants) are available. To have a *meaningful* objective in life (good quality of life) and the ability to use the health resources (*manageability*). These are the three key components of the Sense of Coherence. Comprehensibility is the cognitive component, meaningfulness the motivational component and manageability the behavioural component (see Chaps. 3 and 4).

Twenty years after this Charter some of the key actors involved in the development and implementation of the Ottawa Charter were asked to comment on the development of health promotion over the past 20 years. These reflections on the realization of the Ottawa Charter were published before the 19th IUHPE World Conference on Health Promotion and Education in June 2007 [7, 8]. This was also when the Global Working Group on Salutogenesis was formed. Thirty years later, it is argued that the Ottawa Charter retains its relevance to the present day and that all policy makers and professionals working to promote positive health should revisit and take heed of its principles [9].

The Philosophical Foundation of Health Promotion

The issue of health promotion can be approached from a philosophical point of view exploring theories enhancing health. The biomedical or pathogenic approach where health is generated through the elimination of risks for diseases was the dominating paradigm at that time where the WHO constitution (1948) proclaims health as a three-dimensional concept with physical, social and psychological (mental) dimensions. The definition of Health was manifested and written into the Constitution of WHO which was approved by The UN in 1948 [10]. Scrutinizing the contents of the definition it states that a person who has reached Health in the meaning of the WHO definition is in a *state of complete wellbeing* regarding all three dimensions of wellbeing. Before reaching this absolute level of health one has to move from disease to the absence of disease and furthermore move to a state of complete wellbeing (see Fig. 1.2.). The definition actually excludes all persons with disabilities from the possibility of having a status of Health according to the definition! Health was seen as an endpoint and a static concept not as the dynamic process presented in the Ottawa Charter.

Further, there was a discussion of adding a fourth dimension (the spiritual or existential dimension) where people realized their dreams and ambitions adding

Fig. 1.2 The WHO definition of health as a state of complete well-being. Source: Bengt Lindström

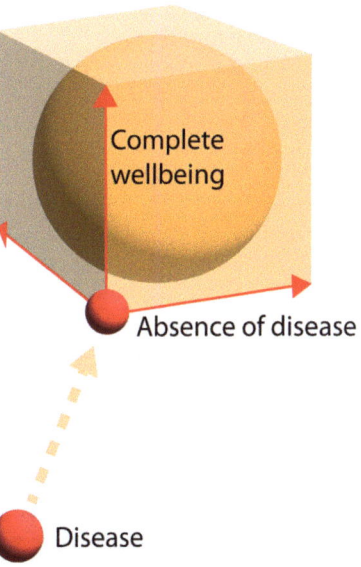

meaning to life. At the inauguration of the Nordic School of Public Health in Gothenburg, Sweden, on August 29, 1987, the WHO Director General Halfdan Mahler gave a speech. After the speech, Bengt Lindström interviewed him about the WHO definition of health. Below is an excerpt from the interview:

> Let me take one of the founders of WHO, who drafted the definition of health as a physical, social and spiritual wellbeing, and adds, it is the absolute level of that. I said but stupid, and he said: Do you remember during the war when I went out in the evening and did not know whether I was coming back in the morning. That is when I had the highest Quality of Life. Because physically, I was ready to scarify myself for my country, socially, I knew I was surrounded by my friends that would take care of my family if I did not come back and spiritually, mentally, I felt a total wellbeing, because I could make such a decision. So even faced with death you may go out and do not know if you are coming back, you can still have the highest quality of life. This having been said, quality of life of course cannot be defined because it is a quality of spiritual dimension of health because it is something that does not necessarily have to do with whether you are physically an invalid you can still be mentally a super human being and you can still feel that life is marvellous. So, for me the important thing with quality of life is never to believe whether your somatic part is functioning. If we do not add that we are socially functioning well and that you spiritually are functioning well. These are the three key variables in my opinion in quality of life physical wellbeing, social wellbeing and spiritual wellbeing and that is how I would define it. If you get to the highest possible level of that you can still be ill and you can still have that kind of wellbeing. (Halfdan Mahler, WHO Director, 1987, video recording).

The four dimensions of health are shown in Fig. 1.3.

In spite of the theoretical similarities between the intentions of the Ottawa Charter and health promotion as related to the theory of Salutogenesis this connection has not yet been thoroughly explored. Health promotion research is based on a wide range of disciplines where the diversity reflects theories of organizational behaviour, sociology, social psychology, psychology, anthropology, education,

Fig. 1.3 The four dimensions of health. Source: Bengt Lindström

economics and political sciences. (see Chap. 9) [11]. However, at the time critics claimed that the Ottawa Charter did not correspond to contemporary health problems, and that the action areas were obsolete and ineffective or even utopian and therefore not operational. On the contrary, time has shown that the Ottawa Charter represented a relevant response to most of the major challenges. The levers of this Charter have been shown to be effective and efficient and can constitute an integrating framework that can be adapted to contextual and scientific changes in public health. It is therefore not time to abandon this Charter, but, on the contrary, to intensify its deployment [12].

In the first Decade after the Ottawa Charter, the lack of theory became an irritating problem for many leaders of health promotion research [11, 13, 14]. On the other hand, the direction and focus on health—not disease—was a clear advantage compared to the time before [13, 15–17].

A Salutogenic Interpretation of the Ottawa Charter

This section starts with Table 1.1 demonstrating some common and different denominators of salutogenesis and the Ottawa Charter.

Health promotion had enormous ambitions and attempted to cover everything. This diversity has made it difficult to make the core things clear. In addition, in the 90s, the lack of theory made health promotion a giant standing on clay feet. If one uses the salutogenic perspective health promotion can be boiled down to a hard core of only four simple things that have to be in place to make health promotion effective. This can be expressed in a logic philosophical formula. First, there is a need for an understanding of

- *health promotion* (HP) as based on the WHO Ottawa Charter (OC),
- *salutogenesis* (SAL) as the process leading towards health,
- *quality of life* (QoL) as the outcome of the whole process.

Table 1.1 Some common and different denominators of salutogenesis and the Ottawa Charter for health promotion. Updated from the first ed. of *The Hitchhikers Guide* [1]. (Reprinted with permission of © Bengt Lindström. All Rights Reserved)

	Salutogenesis	The Ottawa Charter
Prologue	The holocaust	UN declaration of human rights/WHO declaration of health
Reference	*Antonovsky*, A. (1979). Health, stress and, Coping [18]. *Antonovsky*, A. (1987). Unraveling the Mystery of Health [19]. *Mittelmark* et al., 2017, 2022 (second ed.). *The Handbook of Salutogenesis* [20, 21]. Society for Theory and Research on Salutogenesis (*STARS*), IUHPE *global working group on Salutogenesis, Center of Salutogenesis,* University of Zürich	1986
Status	Theory, evidence	Principles, ideology
Fundament	Human Rights i.e. active participating subjects	Human Rights i.e. active participating subject
Focus	Multidimensional and culturally sensitive life orientation The salutogenic question "what creates health?"	Health promotion
Health as	A lifelong learning process	A process
Key concepts	Sense of coherence (SOC) and generalized (GRRs) and specific resistance resources (SRRs). Sense FOR coherence	Health promotion
Resources	Generalized and specific resistance resources	Health determinants
Key mechanism	Ability to use the GRRs and SRRs for the development of a strong SOC	Enable control over the health determinants (empowerment)
Elements	Comprehensibility, manageability, meaningfulness … there are even others, however, must be confirmed by further research	Health determinants, setting, process, active productive life
Approach	Contextual, systems, culture	Settings
Outcome	Perceived good health, mental health and quality of life	A better health, an active, productive and meaningful life
Professional role	Serves as a GRR and/or SRR. Facilitators for the preconditions.	Facilitator, enabling people
Misconception	Only measuring SOC. SOC as a screening instrument to find persons with weak SOC to be subjected to intervention. SOC as a tool for evaluating activities.	Only a risk approach focusing on health behaviour

$$HP(OC) = SAL + QoL$$

The only necessary addition is to build in human rights (HR) as a fundament making the value of the human being an active participating subject as a rule. For children and young people this again would mean an active use of the child convention where both the child and the context are considered as the value base (UN CRC).

$$HP(OC) = (SAL + QoL) \times HR$$

Health promotion and the Ottawa Charter have their fundament in WHO and the UN which at the core are working on the basis of the Declaration of Human Rights. It is further of utmost importance to explicitly include Human Rights as the ethical fundament for both health promotion and salutogenesis. This basis is unquestioned and overrules everything. Further, many times a logic misconception is made thinking that people who are in good health or people who have developed a strong SOC automatically are good human beings. However, there are no morals connected to good health or a strong SOC [22]. There is nothing automatic, it has to be made explicit. To conclude, the simple elements of the salutogenic model for Health Promotion and the Ottawa Charter are resources, meaning, motivation and action competence.

References

1. Lindström, B., & Eriksson, M. (2010). *The hitchhiker's guide to salutogenesis. Salutogenic pathways to health promotion*. The IUHPE Global Working Group on Salutogenesis and Folkhälsan Research Center.
2. Eriksson, M. (2007). *Unravelling the mystery of salutogenesis. The evidence base of the salutogenic research as measured by Antonovsky's sense of coherence scale*. Doctoral thesis, Åbo Akademi University, Folkhälsan.
3. WHO. (1986). *Ottawa Charter for health promotion: an International Conference on Health Promotion, the move towards a new public health*, November 17–21, 1986. World Health Organization.
4. WHO. (1978). *Declaration of Alma-Ata, International Conference on Primary Health Care*, Alma-Ata, USSR, 6–12 September 1978.
5. WHO. (1981). *Global strategy for health for all by the year 2000*. World Health Organization.
6. WHO. (2009). *Milestones in health promotion. Statements from global conferences*. World Health Organization.
7. Hills, M., & McQueen, D. V. (2007). At issue: Two decades of the Ottawa charter. *Promotion & Education, 14*(2 suppl), 5. https://doi.org/10.1177/10253823070140020101x
8. IUHPE. (2007). *The Ottawa charter for health promotion: A critical reflection*. Background to the 19th IUHPE World Conference on Health Promotion and Health Education, "Health Promotion Comes of Age: Research, Policy & Practice for the 21st Century", 14(2 suppl.)
9. Thompson, S. R., Watson, M. C., & Tilford, S. (2018). The Ottawa charter 30 years on: Still an important standard for health promotion. *International Journal of Health Promotion and Education., 56*(2), 73–84. https://doi.org/10.1080/14635240.2017.1415765
10. United Nations (UN). (1948). *Universal declaration of human rights*.
11. Nutbeam, D., Harris, E., & Wise, M. (Eds.). (2010). *Theory in a nutshell. A practical guide to health promotion theories*. McGraw-Hill.
12. Alla, F. (2016). Should the Ottawa charter still be the reference 30 years later? *Santé Publique, 28*(6), 717–720.
13. Kickbusch, I. (2006). The health society: The need for a theory. *Journal of Epidemiology and Community Health, 60*(7), 561.

14. McQueen, D. V., Kickbusch, I., et al. (2007). *Health and modernity. The role of theory in health promotion*. Springer.
15. Morgan, A., & Ziglio, E. (2007). Revitalising the evidence base for public health: An assets model. *IUHPE - Promotion & Education, 2*(Supplement), 17–22.
16. Eriksson, M., & Lindström, B. (2008). A salutogenic interpretation of the Ottawa charter. *Health Promotion International, 23*(2), 190–199.
17. Morgan, A., Davies, M., & Ziglio, E. (2010). *Health assets in a global context: Theory, methods, actions*. Springer.
18. Antonovsky, A. (1979). *Health, stress and coping*. Jossey-Bass.
19. Antonovsky, A. (1987). *Unraveling the mystery of health. How people manage stress and stay well*. Jossey-Bass.
20. Mittelmark, M. B., Sagy, S., Eriksson, M., Bauer, G. F., Pelikan, J. M., Lindström, B., et al. (Eds.). (2017). *The handbook of salutogenesis*. Springer.
21. Mittelmark, M. B., Bauer, G. F., Vaandrager, L., Pelikan, J. M., Sagy, S., Eriksson, M., et al. (2022). *The handbook of salutogenesis* (2nd ed.). Springer.
22. Antonovsky, A. (1995). The moral and the healthy: Identical, overlapping or orthogonal? *Israel Journal of Psychiatry and Related Sciences, 32*(1), 5–13.

Open Access This chapter is licensed under the terms of the Creative Commons Attribution-NonCommercial-NoDerivatives 4.0 International License (http://creativecommons.org/licenses/by-nc-nd/4.0/), which permits any noncommercial use, sharing, distribution and reproduction in any medium or format, as long as you give appropriate credit to the original author(s) and the source, provide a link to the Creative Commons license and indicate if you modified the licensed material. You do not have permission under this license to share adapted material derived from this chapter or parts of it.

The images or other third party material in this chapter are included in the chapter's Creative Commons license, unless indicated otherwise in a credit line to the material. If material is not included in the chapter's Creative Commons license and your intended use is not permitted by statutory regulation or exceeds the permitted use, you will need to obtain permission directly from the copyright holder.

Chapter 2
A Change of Perspective

Bengt Lindström

The Making of Blind Men

Aaron Antonovsky often used metaphors to explain salutogenesis. One of his favourites was the story of "The making of blind men" [1]. This is how it goes: At the age of 35 a man in the prime of life suddenly becomes completely blind. Thereafter his surroundings define him as the blind person. He is treated as a totally disabled and disorientated person. People stopped listening to his opinions and saw him only as "the blind man." There was of course the good intention to help, but he was made "the blind man." Initially, he found it difficult to find his way in the world of the seeing but he learned how to manage by trial and error and with the help of family, friends and professionals. Looking at him with other eyes, he still has maintained most of his capabilities and functions. This is what is important Antonovsky said and turned the question of health around, *saying we should look at what creates health rather than only what the limitations and the causes of disease are.* By raising the question, we are given different answers and find different solutions and identify resources of health.—Instead of making him the blind man he—in salutogenic terms—should be defined according to his capabilities; "Here is a man, a skilled engineer and top manager, a good husband and father. He looks attractive, he is pleasant and well-mannered, but he has the difficulty of sharply distinguishing objects because he is short of seeing."

Another story he used in his lectures is the short story "How to kill a child" written by Stig Dagerman, a Swedish author writing prose and poems about difficult themes and moments in life [2]. This story starts on a beautiful summer morning. A

B. Lindström (✉)
Norwegian University of Science and Technology (NTNU), Trondheim, Norway

Nordic School of Public Health, Gothenburg, Sweden
e-mail: bengtblind@hotmail.com

family discovers they have run out of sugar for coffee at breakfast and send their son across the road to the neighbours to fetch some. At the same time, a young man is driving his car with his beloved to the seashore to enjoy this wonderful day. The child eventually gets killed, accidentally overrun by the car. A grim story with an outcome nobody in the story wanted or expected. Antonovsky wanted to tell us that life can never be controlled completely. In a sense, we must live with this unpredictability. To remain healthy still maintain our ability and trust in life. This is again a part of the salutogenic framework, to be able to deal with uncertainty and chaos however much we would like to believe we are in control of life.

The Original Sample of Israeli Women

The original sample included Israeli women aged 45–54 of Central European birth and focused on problems of adaptation to menopause [3]. The respondents constituted a representative sample of all women in a middle-sized Israeli city meeting the criteria of age and nativity. Of the 287 women in the sample, 77 had been in a Nazi concentration camp during World War II. The remaining 210 women were used as a control group. The detailed data showed camp survivors to be more poorly adjusted than the controls. Of greater importance, however, was the fact that several concentration camp survivors were found to be well-adapted, despite the extreme trauma. Three complementary explanations of the fact of successful adaptation were proposed: an initial underlying strength, a subsequent environment that provided opportunities to reestablish a satisfying and meaningful existence, and a "hardening" process that allows the survivor to view current stresses with some equanimity. The distribution of the original sample is shown in Fig. 2.1.

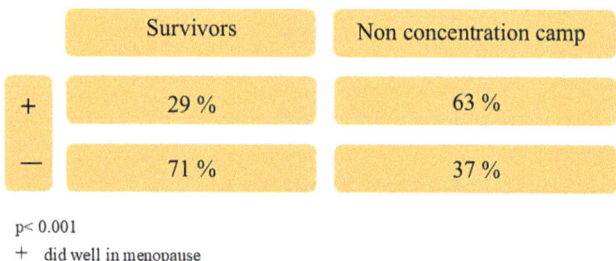

p< 0.001
+ did well in menopause
− problems in menopause

Fig. 2.1 A description of the original sample studying the effects of menopause of Israeli women [3]

How Antonovsky Graphically Described his Theory

His theory was originally aimed to be a theory of stress and stress as a natural part of life. Therefore, the factors that can upset one's position are named *"stressors."* Under the influence of stressors, one comes under tension thereafter either succumbs to stressors leading to a breakdown in a pathogenic direction or overcomes the strain and moves towards the health end of the continuum, i.e. moves in the salutogenic direction. Fig. 2.2 is based on Antonovsky's original way of drawing. He emphasized that most of the time research has been interested in the mechanisms behind breakdown (or the pathogenic orientation) while his focus was on what resources, conditions and factors can make us move in the health direction (a salutogenic orientation).

Antonovsky himself drew the health continuum or as he said the fully appropriate term "the ease/dis-ease continuum" as a horizontal line between total absence of health (H−) and total health (H+) and explained that all people are positioned somewhere on this line [5]. We encounter stressors every day that we have to deal with. Stressors can upset our position, and we come under tension. Here there are two options, either the pathogenic forces overtake us, and we break down or we regain our health through salutogenesis and move towards H+. Conceptually salutogenesis is the direction towards the H+.

One of the keys to the salutogenic approach was to describe *health as a continuum* between "total health and total unhealth" or *ease—dis-ease continuum.* Conceptually Antonovsky defined the breakdown or health ease/dis-ease continuum "as a multifaceted state or condition of the human organism" [5, p. 65]. Each of us can be placed at any time in this continuum. Antonovsky also said that in this continuum one simultaneously can have both healthy and unhealthy components. Health thus becomes more relative than WHO's definition of health (as a complete state of well-being and not only the absence of disease).

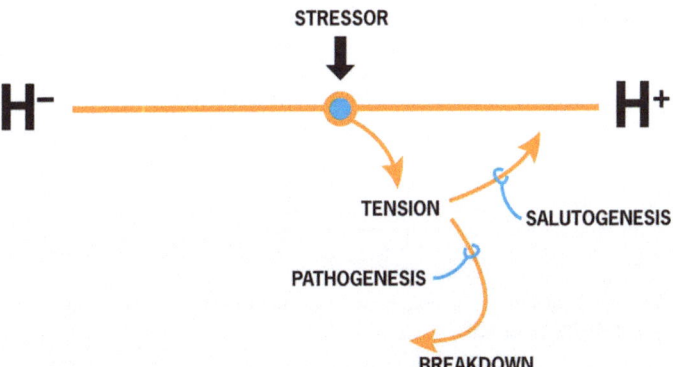

Fig. 2.2 The health continuum "ease/dis-ease" [4]. (Reprinted with permission of © Bengt Lindström. All Rights Reserved)

Antonovsky [5, p. 65] defined the health ease/dis-ease continuum as follows (Fig. 2.3):

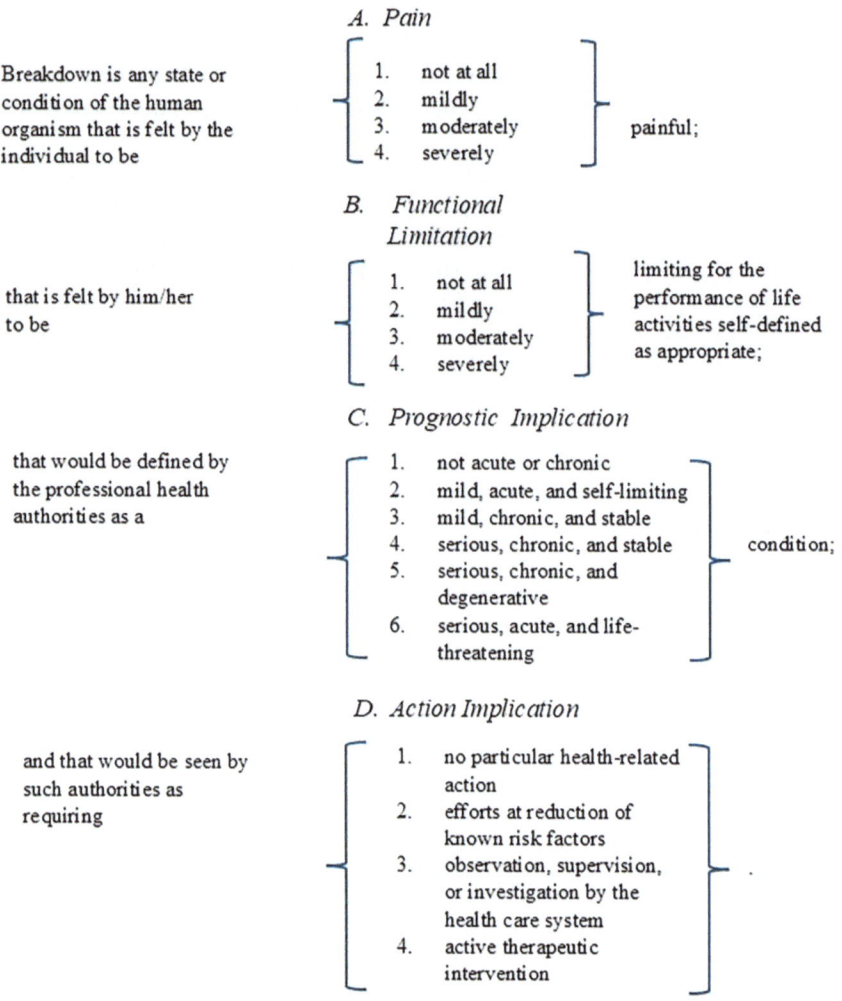

Fig. 2.3 Measuring health on a continuum [5] (Reprinted with permission of © Avishai Antonovsky. All Rights Reserved)

From the River of Health to the River of Life

Traditionally, the difference between the biomedical model and public health has been described metaphorically as a river. The following stages are described moving up the river: (i) cure or treatment of diseases; (ii) health protection/disease prevention; (iii) health education and finally on top health promotion. Health promotion holds a rather different perspective relating mainly to resources for health and life not primarily risk and disease. All approaches ultimately strive to improve health, but from different perspectives. This is a classic image called *The River of Health* (Fig. 2.4) where "the downriver bias" focuses on processes where the risk exposure already may have caused damage (cure, protection, prevention and often health education) [6]. The health concept in this paradigm is constructed from the understanding of disease, illness and risks. However, in the health promotion approach, we bring the focus upstream to finding resources, initiating processes not only for health but also well-being and quality of life. This classic image explains the difference between care, protection, prevention and health education and opens for health promotion. The River of Health is a simple way to demonstrate the scene of actions for health.

In the salutogenic approach, we focus on the direction towards health. The ultimate objective of health promotion activities is to facilitate prerequisites for a good life. Perceived good health is a determinant of quality of life.

Antonovsky did not live long enough to elaborate on these images. In our reading and thinking on salutogenesis, we have changed the river into a different and more salutogenic framework placing Health in the River of Life (Fig. 2.5). Here the main flow of the river is in the direction of life while illness, disease and risks are seen as disruptive forces one will encounter in life—still life as such is the main force and the main direction. Antonovsky explicitly talked about resources for life and constructed a life orientation questionnaire, the sense of coherence (SOC) questionnaire. Here Antonovsky's ease—dis-ease continuum is placed vertically. To explain

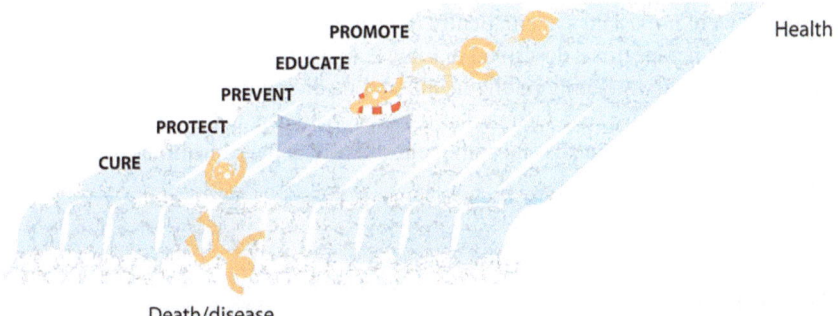

Fig. 2.4 The River of Health [4]. (Reprinted with permission of © Bengt Lindström. All Rights Reserved)

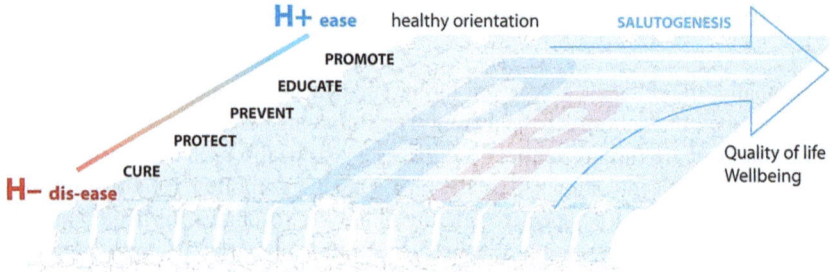

Fig. 2.5 Health in the River of Life [4]. (Reprinted with permission of © Bengt Lindström. All Rights Reserved)

the shift of paradigm of the salutogenic framework, the metaphor of the river needs to be different. This is the River of Life [7]. Here the river flows vertically across your view. Along the front side of the river, there is a continuous waterfall following the whole stretch of the river meaning wherever you are there is always a possibility to encounter risks, disease and death. However, the main flow and direction of the river is not down the waterfall but running vertically in the direction of life.

Some are born *at ease* where the river floats gently, where there is time to learn, where one can float and the prerequisites for life are good with many resources at disposal, like being born in a welfare society. Others are born close to the waterfall, *at dis-ease*, where the struggle for survival is hard and the risk of going over the rim is much greater. The river, just like life, is full of risks and resources, however, our outcome is based on our orientation and learning through our life experiences thus acquiring an ability to identify and use the resources necessary to improve our options for a better health and quality of life.

The health process is a learning process where we reflect on what will create health and what are the options for life and improve QoL. If we never ask these questions, we never know the answers and never learn. Before Antonovsky these questions were not asked systematically in health science.

Note: The perspectives described in Figs. 2.4 and 2.5 are not contradictory. They are complementary meaning public health and medicine can very well use a health promotion perspective and vice versa. It is the synthesis and synergy that will be most effective. Salutogenesis and pathogenesis work at the same time (see Fig. 8.2 in Chap. 8).

References

Scott, A. R. (1981). *The making of blind men. A study of adult socialization*. Transaction Publishers.
Dagerman, S. (1952). *How to kill a child [Att döda ett barn]*. Stockholm.

Antonovsky, A., Maoz, B., Dowty, N., & Wijsenbeek, H. (1971). Twenty-five years later: A limited study of the sequelae of the concentration camp experience. *Social Psychiatry, 6*(4), 186–193. https://doi.org/10.1007/BF00578367

Lindström, B., & Eriksson, M. (2010). *The hitchhiker's guide to salutogenesis. Salutogenic pathways to health promotion*. Folkhälsan, Helsinki, Finland and IUHPE Global Working Group on Salutogenesis.

Antonovsky, A. (1979). *Health, stress and coping*. Jossey-Bass.Antonovsky, A. (1987). *Unraveling the mystery of health. How people manage stress and stay well*. Jossey-Bass.

Eriksson, M., & Lindström, B. (2008). A salutogenic interpretation of the Ottawa Charter. *Health Promotion International, 23*(2), 90–199. https://doi.org/10.1093/heapro/dan014

Open Access This chapter is licensed under the terms of the Creative Commons Attribution-NonCommercial-NoDerivatives 4.0 International License (http://creativecommons.org/licenses/by-nc-nd/4.0/), which permits any noncommercial use, sharing, distribution and reproduction in any medium or format, as long as you give appropriate credit to the original author(s) and the source, provide a link to the Creative Commons license and indicate if you modified the licensed material. You do not have permission under this license to share adapted material derived from this chapter or parts of it.

The images or other third party material in this chapter are included in the chapter's Creative Commons license, unless indicated otherwise in a credit line to the material. If material is not included in the chapter's Creative Commons license and your intended use is not permitted by statutory regulation or exceeds the permitted use, you will need to obtain permission directly from the copyright holder.

Chapter 3
The Original Salutogenic Framework

Monica Eriksson

The Salutogenic Model of Health

The name Salutogenesis stems from the words salus (from Latin, health) and genesis (from Greek, origin) literally meaning the origin of health. In a lecture, "The Salutogenic Approach to Aging," held in Berkeley on January 21st, 1993, Aaron Antonovsky conceptually defined salutogenesis as

> the process of movement toward the health end of a health ease/dis-ease continuum [1].

As a medical sociologist, it was natural for Antonovsky to focus on the human being always interacting with the context, this refers to a system theory thinking. An environment characterized by chaos and constant change, he stated, was a normal state of life, that is a heterostatic view of society as opposed to a homeostatic state [2]. The key elements are the orientation towards problem solving and, the capacity to use the resources available [3].

The key concepts in the salutogenic theory are the *sense of coherence* (SOC, see Chap. 4) and the *generalized* (GRRs) and *specific resistance resources* (SRRs), as extensively explained below. Fundamental in the salutogenic theory is to consider health as a position on a health ease/dis-ease continuum and the movement in the direction towards the health end (see Chap. 2, Fig. 2.1). Antonovsky talked about processes for health strongly rejecting the dichotomy between health and disease. Already in 1993, he described salutogenesis as a theory of health. Today there is consensus around weaving salutogenesis not only as an approach to a good health development but also as a theory, an orientation to life and as a sense of coherence (SOC) [4]. The salutogenic approach focuses on resources for health and health-promoting processes. From the early stage of development of the salutogenic

---

M. Eriksson (✉)
Department of Psychology, Lund University, Lund, Sweden
e-mail: monica.eriksson@psy.lu.se

theory, the intention was to apply the theory at an individual and a group (family) level [5], later also at a society level [6–8]. There is extensive research today at the individual level as well as in groups (families) and especially in the area of workplaces.

Certain trends in salutogenic research are evident: (1) from the early beginning in the 1980s until today, the focus has been on testing the SOC questionnaire in different countries all over the world, using different samples such as the general population, patients, families, parents, children, adults and older people; in various disciplines among others in public health, health promotion, oral health, nursing sciences, in training and educational sciences as well as in learning sciences (see Chap. 7); (2) developing the SOC questionnaire for application on a societal level (see Chaps. 4 and 9); (3) the translation of the SOC questionnaire to languages other than English (see Chap. 4, Fig. 4.2); (4) an increasing focus on the structural validity of the SOC questionnaire (see Chap. 4); (5) an increasing appreciation to consider salutogenesis as a theory of health, wellbeing and quality of life by applying the salutogenic principles in practices (see Chaps. 5 and 6); (6) using salutogenesis as the theoretical framework for interventions aiming at investigate the effectiveness and the potential to strengthen the SOC, e.g. in workplace health promotion (see Chap. 9); and (7) an increased interest to develop new questionnaires to measure salutary factors, i.e. salutogenic questionnaires (see Chap. 3). Today there are also many qualitative studies using different methods for gathering and analysing data.

Sometimes the word salutogenic is confused with salutogenetic, which has another meaning. However, research on that issue is scarce. Hansson and colleagues reported findings from a study on Swedish twin mothers [9]. The Twin Mother's Study examined the influences on maternal adjustment, especially the relative importance of genetic and environmental factors for mental health. They found that 35 percent of the SOC depended on genetic effects and 57 percent depended on non-shared environmental effects [9, p. 544]. Similar results are found among Finnish monozygotic twins at 20–27 years of age, where genetic factors explained 39 percent of the variation of SOC in males and 49 percent in females [10].

Generalized and Specific Resistance Resources

According to Antonovsky [11, p. 103] the GRRs are defined and shown in Fig. 3.1.

The GRRs are the corner stones in the salutogenic theory of health, creating the prerequisites for the development of a strong SOC by providing an individual with sets of life experiences characterized by consistency, participation in shaping outcome and an underload and overload balance [5, p. 19]. They can be found within people as resources bound to their person and capacity but also to their immediate and distant environment as of both material and non-material qualities, from the person to the whole society. Examples of GRRs are money, housing, self-esteem, knowledge, heredity, healthy orientation, contact with inner feelings, social

3 The Original Salutogenic Framework

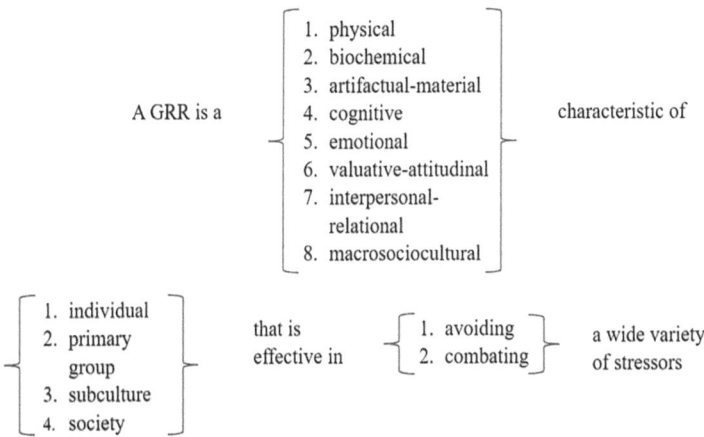

Fig. 3.1 Mapping-sentence definition of a generalized resistance resource [11, p. 103] (Reprinted with permission of © Avishai Antonovsky. All Rights Reserved)

relations, existential issues, beliefs, religion and meaning of life and culture [5, 11]. At least four of the GRRs have to be at one's disposal in order to facilitate the development of a strong SOC: meaningful activities, existential thoughts, contact with inner feelings and social relations [5, p. 23]. These are the GRRs as Antonovsky expressed them in the late 70 s and 80 s. More GRRs can be added, such as cognitive abilities and physical activity [12]; "caregivinghood" among caregivers to older adults [13]; nurses/nursing among cancer patients [14]; salutogenic nursing home care [15]; parental resources, family climate resources, school settings and communities [16]. Further, Griffiths, Ryan and Foster [17] identified several resistance resources to be aware of, among others, structure and predictability in life, future orientation and a positive solution-oriented outlook. The key is not only about having the resources at your disposal but also the ability to use them in a health-promoting way. More recent research highlights the importance of repeated experiences with resources and everyday challenges [18]. The core experience is the ability to re-organize resources and participate in intellectual meaning-making through equal power dialogues. A strong SOC is here described as a deeper understanding of how and why resources work and under which circumstances resources work [18]. This is about learning, not only about health but health and good health development through an internalization of the new knowledge. A certain degree of flexibility is required. Antonovsky [5] discussed flexibility as a potential characteristic of a strong SOC, however, he argued that flexibility even may belong to the manageability dimension.

Where the GRRs are more generic, the *specific resistance resources* are partly tied to the individual and to a specific situation, shown in Fig. 3.2.

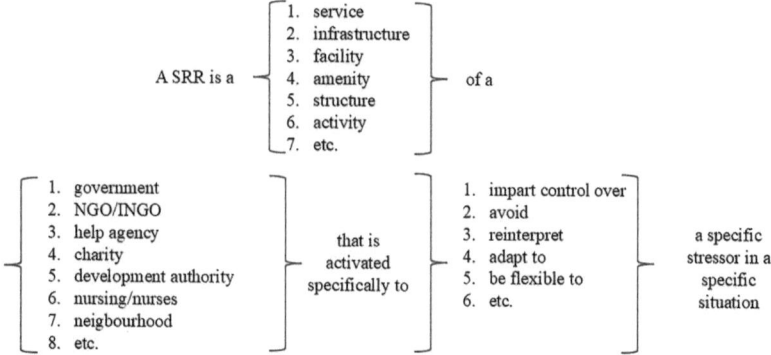

Fig. 3.2 Mapping the sentence definition of specific resistance resources (Adapted with permission from Mittelmark et al., 2017 [19], Fig. 8.2. Some modifications were made. https://doi.org/10.1007/978-3-319-04600-6_8, licensed under the terms of the Creative Commons Attribution-Noncommercial 2.5 License (http://creativecommons.org/licenses/by-nc/2.5/))

Research on the resistance resources after Antonovsky died is limited. There is a need to be aware of and to identify generalized and specific resistance resources both at an individual level but also in the immediate environment. In an ever-changing world, the resistance resources also change. They are crucial for the development of a strong sense of coherence. Most people have dreams about the future, seeks goals and meaning based on their own conditions, often we hear about people who have succeeded. When listening more intensively, it appears that these people have had someone, who believed in them, who saw them as resourceful despite their shortcomings. Such a supporting person is an example of a specific resistance resource, characterized by a promoting view of humanity, such as how Antonovsky talked about "The making of the blind man" (see Chap. 2).

In this book, we consider salutogenesis (SAL) as the theoretical framework for health promotion (HP) practice. If you look more closely at the concepts in the salutogenic theory and health promotion you will find that the content and the principles are the same (see Chap. 1), but the vocabulary may differ between the two approaches [20]. Both are closely connected to each other. Shortly the explanation of Fig. 3.3 is as follows: (1) We are all living in a certain context, called setting in HP, coherent context in SAL. Our life journey is to some extent determined by the resistance resources (GRRs, SRRs) and the health determinants (HD), which we constantly carry with us, here visualized in the form of a backpack. Given the prerequisites for a good health development, we need to understand the context we are part of, need to manage the tension that stressors may cause and at the same time have a pronounced desire to look forward with the help of activities, which give meaning. Life events can be positive as well as negative, however, they all develop

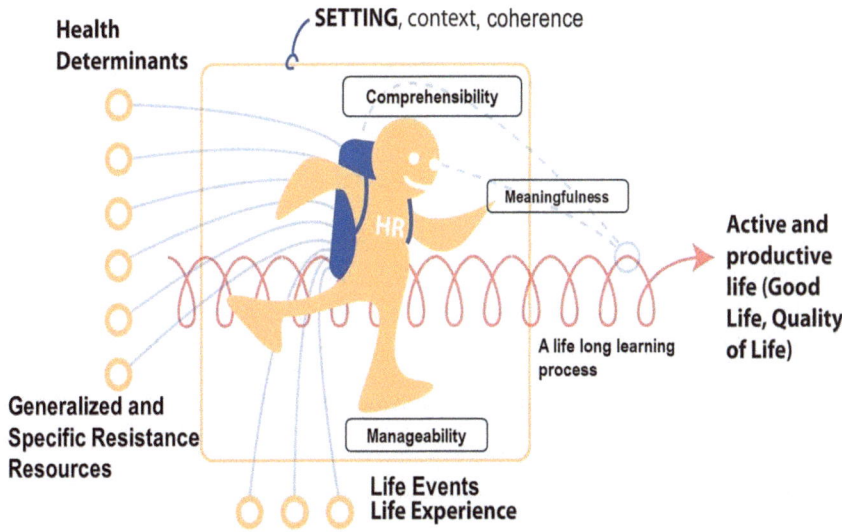

Fig. 3.3 Salutogenesis and the core content of health promotion according to the Ottawa Charter for Health Promotion [20] (Reprinted with permission of © Bengt Lindström. All Rights Reserved)

into life experiences, which should be in the backpack. The life experiences can be used to solve problems and manage stress in the future. The outcome of this journey is a good life and quality of life. A productive life does not mean that we always and everywhere need to be on top of our potential, it means that we live an active and productive life dependent on the prerequisite of ourselves, at a certain stage in life and in a certain life situation. That is enough. Finally, and the innermost core of HP, is the human rights, the ethics (HR), which always has to be in place when meeting and treating people.

The Role of Culture

The concept sense of coherence (SOC) is frequently understood as a cross-cultural concept, meaning that it works in the same way independent of culture. However, culture seems to define which resources are appropriate at all stages of coping with a stressor [21]. Antonovsky [5] stated that how people respond to stressors and the choice of resources at disposal they use are shaped by one's culture. From the very beginning, he has had culture in mind: "certain cultural and historical situations foster a strong sense of coherence" [11, p. 149], especially cultures characterized by cultural stability. "Subculture and cultural patterns of social organization provide continued series of experiences that build up the GRRs, crucial to a strong sense of coherence" [11, p. 152]. He continued, "the actual content of behavior, of the resources chosen to cope with stressors, is always shaped by one's culture." The concept may be cross-cultural, but its concrete translation will vary widely; ... the confident may be a relative of the older generation, a holy person, a spouse, or a friend... one's culture defines which resources are appropriate and legitimate in a given situation [5, p. 148]. These statements are important for understanding GRRs and SRRs of today. For further understanding of the potential impact of culture on SOC see Chap. 4, describing more recent research on the structural validity of the SOC questionnaire.

In a review of the role of culture in salutogenesis, Benz et al. [22, p. 16] found that Antonovsky's expansive interest in the role of culture was focused on "shaping life situations, giving rise to stressors and resources, contributing to life experiences of predictability, load balance and meaningful roles, facilitating the development of the sense of coherence and finally shaping perceptions of health and wellbeing". Antonovsky viewed culture as an integral part of the salutogenic theory from the early beginning.

Salutogenesis as an Umbrella Concept

Salutogenesis is much more than only the measurement of SOC. It is a resource-oriented approach focusing on health and how people can manage stressors in daily life. It is a broad concept focusing on resources, competencies, abilities and assets on different levels, the individual, the group such as families and workplaces, and in societies.

Salutogenesis represents an umbrella concept consisting of many different theories and concepts with the focus on resources at different levels, generalized as well as specific resistance resources, all essential and fundamental for explaining health and wellbeing from a salutogenic perspective. Research on vital theories and concepts is limited. One exception is an anthology Health Promotion in Health Care—Vital Theories and Research [23], as far as we know the only book that highlights salutogenic theories and concepts, relevant to the healthcare sector. Several of these concepts are visualized in Fig. 3.4.

There are similarities and differences between the concepts in the figure, however, they all have elements and dimensions that can be considered as salutogenic. Their focus is on resources for good health development and an experience of quality of life. WHO uses the term asset approach while we in this book use resource approach [24]. Next, some of the concepts related to SOC are scrutinized, trying to answer the question of how these related concepts fit into the salutogenic model of health.

SALUTOGENESIS
Theoretical concepts relevant to nursing in health care

© Monica Eriksson 2024

Social support | Empowerment | Flourishing | Sense of Coherence | Dignity | Belonging
Self-efficacy | Self-transcendence | Hope | Will to meaning | Willpower | Connectedness
Salutogenic nursing | Nurse-patient interaction | Person-centered care | Inner strength
Bodyknowledging | Coping

Reasonableness | Resilience | Learned resourcefulness | Attachment | Empathy | Wellbeing
| Learned hopefulness | Humour | Gratitude | Quality of Life | Flow | Hardiness | Social capital
Locus of Control | Ecological system theory | Interdisciplinarity | Cultural capital | Thriving
Posttraumatic Personal Growth | Learned optimism | Slow nursing | Self-tuning

Fig. 3.4 Salutogenic theories and concepts relevant to health care [20]. Modified after Lindström and Eriksson, *The Hitchhiker's Guide to Salutogenesis*, 2010. Reproduced with permission from © Monica Eriksson. All Rights Reserved)

Resilience

We begin with the concept of resilience (resiliency), to many a familiar concept related to coping with adversities, and which is usually considered the same as SOC [25–28]. Do resilience and SOC describe the same phenomenon? Antonovsky met Emmy Werner in the beginning of the 1980s, and then had the opportunity to discuss how the two concepts may be related to each other [5]. In the book *Vulnerable but Invisible,* about the children on the island of Kauai [25, p. 154], is stated "… a more internal locus of control, a more positive self-concept, and a more nurturant, responsible, and achievement-oriented attitude toward life … a sense of coherence in their lives …". In our opinion the answer is no, however, there are similarities and differences between the two concepts [29]. First, the starting points are completely different. While Antonovsky refers to a positive outcome independent of stress under certain conditions, research on resilience starts by recognizing the risk of a negative health outcome. Both the concepts are multidimensional, and process oriented in a continuum, not part of personal characteristics. The salutogenic theory describes generalized resistance resources, factors that create the prerequisites for the development of a strong SOC. Resilience research follows a similar reasoning but names them protective factors for a positive health development. The first one focuses on health promotion, while the second one emphasizes health protection. Both the SOC concept and resilience can be applied to different levels; the individual, group (families) or society. The measurement of the two concepts differs. SOC can be measured through one of the original scales (SOC-29, SOC-13) or some modified translations, modified in the sense that the items are labelled in the same ways, but the scoring alternatives differ as well as the number of items included in the studies. Depending on which level of resilience is studied there is a need to use different scales with different items and scoring alternatives.

More recent research states that "resilience and salutogenesis address different research problems at slightly different systemic levels. Suffice it to note for the moment that the two fields touch, and even overlap, but they remain distinctive" [30, p. 153]. According to Mittelmark resilience has three components; exposure to significant adversity, a set of behaviours that signal coping and a set of multilevel processes that result in degrees of coping [30, p. 154]. In an integrative review of international literature on mental health nursing, Foster et al. [31] describe studies dealing with the theoretical conceptualizations of resilience. They sum up that resilience can be seen as primarily an individual ability or characteristic, as an interactive person-environment process, and as a collective capacity. This is partly an opposite view of Antonovsky, who was particularly clear that SOC was not seen neither as a personality trait or a personality type, nor a coping strategy, but as a life orientation, where the three dimensions, comprehensibility, manageability and meaningfulness, interacted depending on the situation [32, p. 37]. Métais et al. [33] call for clarification of the definitions of resilience. They reviewed 69 papers and found that resilience nowadays is about "adapting and bouncing back to previous levels of health" or about "thriving and rising above the adversity towards increased levels of health."

Empowerment

Although the Brazilian education scientist Paulo Freire never used the concept of empowerment, he still is the person who symbolizes the shift of education practice towards empowerment in the sense of making learning available to all, especially to the underprivileged and oppressed [34]. He aimed to reduce inequity through this learning process and mobilize the uneducated. The core is centred on the creation of a respectful dialogue thereby enhancing a sense of social community, i.e. building social capital. In a recently published anthology of salutogenic concepts, among others empowerment, Tveiten states that the concept of empowerment is a broad and conceptually complex construct, defined and explained in different ways depending on where the focus is; at an individual, a group or a societal level [35]. In this chapter, the concept of empowerment is shortly highlighted from a health, quality of life and well-being perspective [36].

According to The Health Promotion Glossary of Terms 2021 [37, p. 14] empowerment is defined as "a process through which people gain greater control over decisions and actions affecting their health." Furthermore, "a distinction is made between individual and community empowerment, where individual empowerment refers primarily to the individuals' ability to make decisions and have control over their personal health decisions. Community empowerment involves individuals acting collectively to gain greater influence and control over the factors shaping the determinants of health in their community and is an important goal in community action for health. These concepts are linked and reciprocal. Empowered individuals create empowered communities, and vice-versa." [37, p.14] This is in line with the Ottawa Charter for Health Promotion, the core policy document of WHO [36]. Empowerment is about giving people control and mastery over their lives similar to the enabling process, which focuses on the positive, dynamic and empowering aspects of health [38]. However, in thematic synthesis, Halvorsen et al. conceptually explored empowerment in healthcare [39, p. 1270]. They found that empowerment was viewed as a helping process rather than redistribution of power and that the user perspective often seemed limited.

Research on the relationship between empowerment and SOC is limited. Koelen and Lindström [40] conceptually discussed the role of empowerment in health promotion based on the salutogenic framework. They concluded that empowerment is still more seen as a principle or an idea rather than a solid theory. In hospitals, central principles of empowerment are the distribution of power from the health professionals to the patients, patient participation and acknowledging the patient as an expert regarding herself/himself [35, p. 167]. However, the dimensions of the SOC are closely related to the principles of empowerment (see Chap. 4), but the concept of empowerment is still a distinct concept from the SOC. Klepp et al. argue that there are an additive, overlapping and interactive relation between empowerment and SOC [41].

What Does it Mean to Be Salutogenic?

In a public health conference in Gothenburg Antonovsky expressed some salutogenic words of wisdom to the audience by saying, "... *think salutogenically and act salutogenically*" [42]. The immediate question then becomes, what does it really mean to think salutogenically, to be salutogenic and to act salutogenically? Research focusing on these issues, as far as we know, is limited. This is a task for further qualitative research.

Some new constructs in such a direction can be identified. Investigating community nursing in Norway, Vinje [43, p. 9] raised the salutogenic questions of "How and why do nurses in community health care experience job engagement and stay healthy?" and "Why do they thrive despite adversity?" The answers were found in the development of The Self-Tuning Model of self care [43, 44]. Essential in the model is nurses' capacity for introspection and reflection with deep attention to meaning, meaningfulness and values connected to work challenges, developing a *salutogenic capacity* called *"self-tuning"* [45]. Job engagement was achieved by searching for meaning, the experience of meaning and holding on to meaning as the force of a drive [43, p. 10]. Its fundament was the talent and habit of introspection and reflection.

More recent research exploring why people use complementary and alternative medicine (CAM), introduces the concept of a "salutogenic gaze," an important dimension in the process of interaction in the practitioner-client dyad. Clients in

CAM are not submitting themselves to a cure for a specific disease but seeking expert guidance to achieve better health [46]. Aaron Antonovsky personally in his presentations of the salutogenic theory always pointed out the importance of making sense, but what does it mean to have an ability of sense making? One answer can be found outside the salutogenic research area, in research using The Making Sense Scale among patients with multiple sclerosis [47, p. 97]. Participants who reported having a religious-spiritual belief were more likely to report sense making than those who did not have such a belief. Sense making was related to lower disability and disease severity and evidenced beneficial direct effects on positive adjustment outcomes and depression after controlling for illness and religious-spiritual belief. Further, the concept Sense for Coherence is introduced in Chap. 7.

Concluding Remarks

To sum up, a growing body of research and an increased interest in the salutogenic approach can be seen. Today the salutogenic theory and its core concepts are applied to different samples, on various ages, on general populations to patients affected by various diseases. It is applied in the health sector as well as outside public health and health promotion, such as described here as learning processes. The SOC questionnaire has been used in many countries and different cultures, all over the world. Recent research highlights the role of culture in the salutogenic theory. Continuous research is needed, constantly asking the salutogenic question, "What creates health" and "What does it mean to be salutogenic and to think in a salutogenic way."

References

1. Antonovsky, A. (1993). The salutogenic approach to aging. Lecture held in Berkeley, January 21, 1993.
2. Antonovsky, A. (1985). The life cycle, mental health and the sense of coherence. *Israel Journal of Psychiatry & Related Sciences, 22*(4), 273–280.
3. Lindström, B., & Eriksson, M. (2005). Salutogenesis. *Journal of Epidemiology and Community Health, 59*, 440–442. https://doi.org/10.1136/jech.2005.034777
4. Mittelmark, M. B., & Bauer, G. F. (2022). Salutogenesis as a theory, as an orientation and as the sense of coherence. In M. B. Mittelmark, G. F. Bauer, L. Vaandrager, J. M. Pelikan, S. Sagy, M. Eriksson, et al. (Eds.), *The handbook of salutogenesis* (2nd ed., pp. 11–17). Springer.
5. Antonovsky, A. (1987). *Unraveling the mystery of health. How people manage stress and stay well*. Jossey-Bass.
6. Mana, A., Sagy, S., & Srour, A. (2016). Sense of community coherence and inter-religious relations. *The Journal of Social Psychology, 156*(5), 469–482. https://doi.org/10.1080/00224545.2015.1129302
7. Mana, A., Srour, A., & Sagy, S. (2021). Sense of community coherence, perceptions of collective narratives, and identity strategies in intra- and interreligious group conflicts. *Peace and Conflict: Journal of Peace Psychology, 27*(4), 669–673. https://doi.org/10.1037/pac0000538

8. Abu-Kaf, S., Al-Said, K., & Braun-Lewensohn, O. (2021). Community coherence and acculturation strategies among refugee adolescents: How do they explain mental-health symptoms? *Comprehensive Psychiatry, 106*, 152227. https://doi.org/10.1016/j.comppsych.2021.152227
9. Hansson, K., Cederblad, M., Lichtenstein, P., Reiss, D., Pedersen, N., Neiderhiser, J., et al. (2008). Individual resilience factors from a genetic perspective: Results from a Twin Study. *Family Process, 47*(4), 537–551.
10. Silventoinen, K., Vuoksimaa, E., Volanen, S.-M., Palviainen, T., Rose, R. J., Suominen, S., & Kaprio, J. (2022). The genetic background of the associations between sense of coherence and mental health, self-esteem and personality. *Social Psychiatry and Psychiatric Epidemiology, 57*, 423–433. https://doi.org/10.1007/s00127-021-02098-6
11. Antonovsky, A. (1979). *Health, stress and coping.* Jossey-Bass.
12. Read, S., Aunola, K., Feldt, T., Leinonen, R., & Ruoppila, I. (2005). The relationship between generalized resistance resources, sense of coherence, and health among Finnish people aged 65–69. *European Psychologist, 10*(3), 244–253. https://doi.org/10.1027/1016-9040.10.3.244
13. Wennerberg, M. M. T., Lundgren, S. M., Eriksson, M., & Danielson, E. (2019). Me and you in caregivinghood—Dyadic resistance resources and deficits out of the informal caregiver's perspective. *Aging & Mental Health, 23*(8), 1041–1048.
14. Kvåle, K., & Synnes, O. (2013). Understanding cancer patients' reflections on good nursing care in light of Antonovsky's theory. *European Journal of Oncology Nursing, 17*, 814819. https://doi.org/10.1016/j.ejon.2013.07.003
15. Drageset, S., Ellingsen, S., & Haugan, G. (2023). Salutogenic nursing home care: Antonovsky's salutogenic health theory as a guide to wellbeing. *Health Promotion International, 38*, 1–11. https://doi.org/10.1093/heapro/daad017
16. Idan, O., Eriksson, M., & Al-Yagon, M. (2022). Generalized resistance resources in the salutogenic model of health. In M. B. Mittelmark, G. F. Bauer, L. Vaandrager, J. M. Pelikan, S. Sagy, M. Eriksson, et al. (Eds.), *The handbook of salutogenesis* (2nd ed., pp. 93–105). Springer.
17. Griffiths, C. A., Ryan, P., & Foster, J. H. (2011). Thematic analysis of Antonovsky's sense of coherence theory. *Scandinavian Journal of Psychology, 52*, 168–173. https://doi.org/10.1111/j.1467-9450.2010.00838.x
18. Maass, R. E. K., Lindström, B., & Lillefjell, M. (2017). Neighborhood-resources for the development of a strong SOC and importance of understanding why and how resources work: A grounded theory approach. *BMC Public Health, 17*(1), 704–717. https://doi.org/10.1186/s12889-017-4705-x
19. Mittelmark, M. B., Bull, T., Daniel, M., & Urke, H. (2017). Specific resistance resources in the salutogenic model of health. In M. B. Mittelmark, S. Sagy, M. Eriksson, G. Bauer, J. M. Pelikan, B. Lindström, & G. A. Espnes (Eds.), *The handbook of salutogenesis* (pp. 71–76). Springer International Publishing.
20. Lindström, B., & Eriksson, M. (2010). *The hitchhiker's guide to salutogenesis. Salutogenic pathways to health promotion.* The IUHPE Global Working Group on Salutogenesis and Folkhälsan.
21. Braun-Lewensohn, O., & Sagy, S. (2011). Salutogenesis and culture: Personal and community sense of coherence among adolescents belonging to three different cultural groups. *International Review of Psychiatry, 23*(6), 533–541. https://doi.org/10.3109/09540261.2011.637905
22. Benz, C., Bull, T., Mittelmark, M., & Vaandrager, L. (2014). Culture in salutogenesis. The scholarship of Aaron Antonovsky. *Global Health Promotion, 21*(4), 16–23. https://doi.org/10.1177/1757975914528550
23. Haugan, G., & Eriksson, M. (Eds.). (2021). *Health promotion in health care—Vital theories and research.* Springer.
24. Morgan, A., & Ziglio, E. (2007). Revitalising the evidence base for public health: An assets model. *IUHPE—Promotion & Education, 2*(Suppl), 17–22. https://doi.org/10.1177/10253823070140020701x

25. Werner, E., & Smith, R. (1982). *Vulnerable but invincible. A longitudinal study of resilient children and youth.* McGraw Hill.
26. Werner, E., & Smith, R. (2001). *Journeys from childhood to midlife. Risk, resilience, and recovery.* Cornell University Press.
27. Rutter, M. (1985). Resiliency in the face of adversity. *British Journal of Psychiatry, 147*, 598–611. https://doi.org/10.1192/bjp.147.6.598
28. Luthar, S. S. (Ed.). (2003). *Resilience and vulnerability. Adaptation in the context of childhood adversities.* Cambridge University Press.
29. Eriksson, M., & Lindström, B. (2011). Life is more than survival: Exploring links between Antonovsky's salutogenic theory and the concept of resilience. In K. M. Gow & M. J. Celinski (Eds.), *Wayfinding through life's challenges. Coping and survival* (pp. 31–46). Nova Science Publishers.
30. Mittelmark, M. B. (2021). Resilience in the salutogenic model of health. In M. Ungar (Ed.), *Multisystemic resilience. Adaptation and transformation in contexts of change* (pp. 153–164). Oxford University Press.
31. Foster, K., Roche, M., Delgado, C., Cuzzillo, G., Giandinoto, J., & Furness, T. (2019). Resilience and mental health nursing: An integrative review of international literature. *Internation Journal of Mental Health Nursing, 28*(1), 71–85. https://doi.org/10.1111/inm.12548
32. Antonovsky, A. (1992). Can attitudes contribute to health? *Advances, The Journal of Mind-Body Health, 8*(4), 33–49.
33. Métais, C., Burel, N., Gillham, J. E., Tarquinio, C., & Martin-Krumm, C. (2022). Integrative review of the recent literature on human resilience: From concepts, theories, and discussions towards a complex understanding. *Europe's Journal of Psychology, 18*(1), 98–119. https://doi.org/10.5964/ejop.2251
34. Freire, P. (1970). *Pedagogy of the oppressed.* Penguin.
35. Tveiten, S. (2021). Empowerment and health promotion in hospitals. In G. Haugan & M. Eriksson (Eds.), *Health promotion in health care—Vital theories and research* (pp. 159–170). Springer.
36. Eriksson, M. (2021). Salutogenesis. In F. Maggino (Ed.), *Encyclopedia of quality of life and well-being research.* Springer. https://doi.org/10.10071/978-3-319-69909-7_3445-2
37. WHO. (2021). *The health promotion glossary of terms 2021.* World Health Organization.
38. WHO. (1986). *Ottawa charter for health promotion: An international conference on health promotion, the move towards a new public health*, November 17-21, 1986. World Health Organization.
39. Halvorsen, K., Dihle, A., Hansen, C., Nordhaug, M., et al. (2020). Empowerment in healthcare: A thematic synthesis and critical discussion of concept analyses of empowerment. *Patient Education and Counseling, 103*, 1263–1271. https://doi.org/10.1016/j.pec.2020.02.017
40. Koelen, M. A., & Lindström, B. (2005). Making healthy choices easy choices: The role of empowerment. *European Journal of Clinical Nutrition, 59*(suppl 1), 10–16. https://doi.org/10.1038/sj.ejcn.1602168
41. Klepp, O. M., & Sørensen, T. (2007). Empowerment: Additive, overlapping and interactive relation to sense of coherence, with regard to mental health and its promotion. *International Journal of Mental Health, 9*(3), 5–26. https://doi.org/10.1080/14623730.2007.9721839
42. Antonovsky, A. (1993). *Some salutogenic words of wisdom to the conferees.* Lecture held at The Nordic School of Public Health in Gothenburg, Sweden. Retrieved October 25, 2024, from http://www.angelfire.com/ok/soc/agoteborg.html
43. Vinje, H. F. (2007). *Thriving despite adversity: Job engagement and self-care among community nurses.* Doctoral thesis, University of Bergen.
44. Bakibinga, P., Vinje, H. F., & Mittelmark, M. B. (2012). Self-tuning for job engagement: Ugandan nurses' self-care strategies in coping with work stress. *International Journal of Mental Health Promotion, 14*(1), 3–12. https://doi.org/10.1080/14623730.2012.682754

45. Vinje, H. F., Ausland, L. H., & Langeland, E. (2017). The application of salutogenesis in the training of health professionals. In M. B. Mittelmark, S. Sagy, M. Eriksson, G. F. Bauer, J. M. Pelikan, B. Lindström, et al. (Eds.), *The handbook of salutogenesis* (pp. 307–318). Springer.
46. Brosnan, C., Tickner, C., Davies, K., Heinsch, M., Steel, A., & Vuolanto, P. (2023). The salutogenic gaze: Theorising the practitioner role in complementary and alternative medicine consultations. *Sociology of Health & Illness, 45*(5), 1008–1027. https://doi.org/10.1111/1467-9566.13629
47. Pakenham, K. (2008). Making sense of illness or disability: The nature of sense making in multiple sclerosis (MS). *Journal of Health Psychology, 13*(1), 93–105. https://doi.org/10.1177/1359105307084315

Open Access This chapter is licensed under the terms of the Creative Commons Attribution-NonCommercial-NoDerivatives 4.0 International License (http://creativecommons.org/licenses/by-nc-nd/4.0/), which permits any noncommercial use, sharing, distribution and reproduction in any medium or format, as long as you give appropriate credit to the original author(s) and the source, provide a link to the Creative Commons license and indicate if you modified the licensed material. You do not have permission under this license to share adapted material derived from this chapter or parts of it.

The images or other third party material in this chapter are included in the chapter's Creative Commons license, unless indicated otherwise in a credit line to the material. If material is not included in the chapter's Creative Commons license and your intended use is not permitted by statutory regulation or exceeds the permitted use, you will need to obtain permission directly from the copyright holder.

Chapter 4
The Orientation to Life Questionnaire: Sense of Coherence

Monica Eriksson

The Original Definition of the Sense of Coherence

The core key concept of the salutogenic theory is the Sense of Coherence (SOC). It is defined as

> a global orientation that expresses the extent to which one has a pervasive, enduring though dynamic feeling of confidence that (1) the stimuli from one's internal and external environments in the course of living are structured, predictable, and explicable; (2) the resources are available to one to meet the demands posed by these stimuli; and (3) these demands are challenges, worthy of investment and engagement [1, p. 19].

According to Antonovsky [1] three dimensions form the SOC, that is, the comprehensibility, the manageability and the meaningfulness. Having a strong SOC enables people to view life as coherent, comprehensible, manageable and meaningful, giving an inner trust and confidence to identify resources within oneself and in the immediate environment, an ability to use and reuse these resources in a health-promoting manner. Further, the life orientation (SOC) is a way of thinking, being and acting as a human being, giving the direction of life. It is not only a question about the individual but also the person in interaction with the living context. All the three dimensions interact with each other and are of varying importance in different situations (see Fig. 4.1).

Even if the motivational dimension of meaningfulness sometimes has been seen as the most important [3], research has shown that the value of the different dimensions varies depending on the existing situation. This was the reason why Antonovsky recommended that one should not measure the three dimensions as separate constructs but as a general factor. Several researchers after Antonovsky have explored the dimensions together with SOC and results supported the idea of SOC as a

M. Eriksson (✉)
Department of Psychology, Lund University, Lund, Sweden
e-mail: monica.eriksson@psy.lu.se

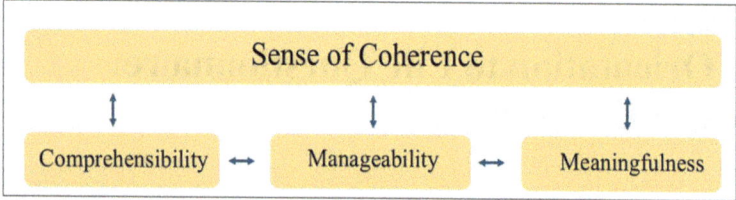

Fig. 4.1 The structure of the original SOC scale with one general factor and three dimensions first published in the first edition of *The Hitchhiker's Guide to Salutogenesis* [2] (Reprinted with permission of © Bengt Lindström. All Rights Reserved)

general factor [4–7]. There are also other results reported. Contrary to Antonovsky [1], recent research suggests SOC to be a multidimensional construct rather than unidimensional [8–10]. According to Antonovsky [11] the SOC is not considered as a coping strategy nor as a personal trait. The SOC can be seen as a coping resource giving people the ability to choose different strategies for solving different problems or to manage life events.

Measuring Sense of Coherence

Originally Aaron Antonovsky developed the Orientation to Life Questionnaire consisting of 29 items to measure the sense of coherence (SOC) on three dimensions of health, the comprehensibility, manageability and meaningfulness [1]. The first step in developing the SOC questionnaire took the form of a qualitative study based on 50 life histories, and open-ended interviews with a variety of people referred by various resources. Included were persons who were known to have undergone severe stressors (concentration camp, lost eyesight, early widowhood…). Interview protocols were read by three colleagues and Antonovsky himself. Each of them classified the respondents on a 10-point scale of weak to strong SOC. They searched for the elements in the way one looked at life and the world. The original questionnaires were developed using facet design, consisting of modality, source, demand (subject) and time, as shown in Fig. 4.2.

Later on a short form, the SOC-13, was introduced [1]. These two questionnaires are the original ones. The summed index of the SOC-29 ranges from 29 to 203 points and 13–91 points for the SOC-13. Most of the studies have used one of these two original questionnaires.

Until 2003 at least 15 modified versions of the SOC questionnaire, ranging from 3 items to 28 items, were found [8]. The questions in the modified versions of the SOC scale are the same as in the original questionnaires, but the scoring alternatives and the number of items included varied. Thereafter, the picture mainly remains, that is, the original questionnaires, especially the short (SOC-13) form, are the most

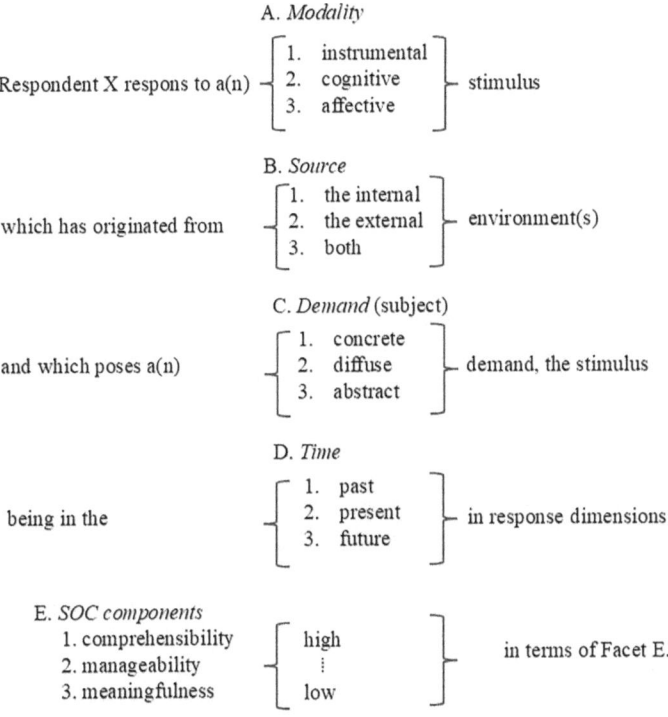

Fig. 4.2 Sense of Coherence Mapping Sentence for Questionnaire Design [1, p. 77] (Reprinted with permission of © Avishai Antonovsky. All rights reserved)

used. Up to date, there are additional modifications; SOC-3 [12], Work-SOC-9 [13], the Leipzig short form, SOC-L9 [14–16]; SOC-12 [17], SOC-13 (5-point) [18], SOC-8 [19], The University of Tokyo Health Sociology version (SOC-3-UTHS) [20] and an abbreviated 5-item version of the original SOC-13 questionnaire [21].

There is a SOC questionnaire adjusted for children (CSOC) [22, 23]; for families (FSOC) [24–26]; for workplaces (Work-SoC Questionnaire) [27, see Chap. 8], for people with limited capability for work, the Sustainable Employability Questionnaire [28] and for societies, the Sense of Community Coherence (SOCC) [29]. The latter should not be confused with the child SOC.

The SOC questionnaire measures how people can manage stress in different life situations. It is not a diagnostic or screening instrument nor a scale evaluating varying activities. Having a weak SOC is not a diagnosis tied to the individual, the SOC depends on the generalized and specific resistance resources. The use of the SOC questionnaire is subject to copyright by Dr. Avishai Antonovsky. Permission to use it can be obtained from The Society for Theory and Research on Salutogenesis (STARS) by application at www.stars-society.org.

The Distribution of the SOC Questionnaires

A systematic review showed that as of 2003, the SOC questionnaires had been used in at least 33 different languages in 32 different countries [8]. A later update showed that even more countries had used the questionnaires, at least in 51 different countries around the world [30]. An update per September 2024 shows further expansion: Andorra [31], Argentina and Malta [32], Cambodia, Hong Kong, Indonesia, Laos, Malaysia, Vietnam [33], Ecuador [34], Chile [35], Ghana [36], Lebanon [37], Morocco [38], Namibia and Puerto Rico [39], Pakistan [40], Philippines [41], Saudi Arabia [42] and United Arab Emirates [43]. In sum, the SOC questionnaires have been used in at least 73 countries all over the world. The distribution of studies using the SOC questionnaires per September 2024 is shown in Fig. 4.3.

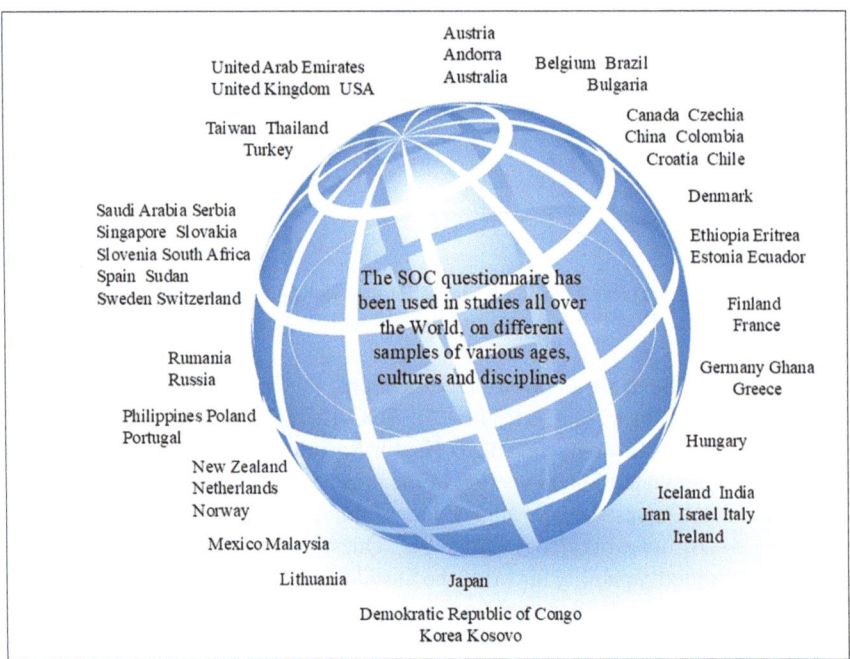

Fig. 4.3 The distribution of studies using the sense of coherence scale 1992–2024 in a global context. Updated and revised after Fig. 8 in the first edition of *The Hitchhiker's Guide to Salutogenesis* [2]. (Reprinted with permission of © Monica Eriksson. All Rights Reserved)

The Dimensionality of the SOC

Antonvosky [1, 44] stated that the SOC scale can be seen as unidimensional, consisting of one general factor (sense of coherence) with three dimensions comprehensibility, manageability and meaningfulness, all three closely related to and interacting with each other (see Fig. 4.1). More dimensions, flexibility, trust and harmony, were discussed. However, discussions ended up with the solution, that flexibility and trust could be included in the manageability whilst harmony could be seen as belonging to meaningfulness dimension. Research focusing on the structure of the SOC scale has attracted increasing interest in recent years. Today there are several studies exploring the structure using confirmatory factor analysis (CFA), resulting in a varying number of models [42, 45, 46]. Research on the structure of the scale has in some studies supported a unidimensional structure [20, 47, 48], others that the scale seems to be a multidimensional construct rather than just one general factor with three dimensions [18, 49].

The SOC questionnaire seems to be culture sensitive. While existing literature demonstrated the cultural relevance of SOC-13 across various countries, there is a study that reflected the cultural differences in several items on the Chinese translation of the SOC-13 [50]. It seems some questions do not make sense for older adults. A translation is not only about linguistics but also of deep understanding of lived experiences of the respondents. There are several studies that report problems in understanding the meaning of different items in the SOC questionnaire. In the Chinese culture, there are some concepts that may explain some of the differences. In a systematic review of differences and common ground in the frameworks of HQoL in traditional Chinese medicine Ding et al. [50] highlight three key concepts: unity of body and spirit ("xingshentongyi"), harmony between man and nature ("tianrenheyi") and seven emotional forms ("qiqing"). Knowledge about these central concepts in the Chinese culture would probably help to further develop the SOC questionnaire. In addition to the core concept SOC in the salutogenic theory of health, the generalized and specific resistance resources (GRRs, SRRs), factors essential for the development of a strong SOC and for maintenance of life challenges, all of this must be culturally identified and explored.

Stability Over Time

Antonovsky assumed the total SOC develops until 30 years of age, then remains stable until retirement and thereafter decreases. This assumption has not been empirically supported, instead, the SOC seems to be relatively stable over time. However, not as stable as Antonovsky presumed [8, 51–53]. SOC seems to become more stable over age [49], suggesting there is a slight increase over time throughout the lifespan. Since we are talking about life orientation the SOC is a rather stable entity. Therefore, it is not surprising the SOC is quite stable and resumes its stability after stress as compared to other short-term phenomena. An overview of studies exploring temporal stability over time is shown in Table 4.1.

Table 4.1 Observational studies that report longitudinal changes in SOC among adult populations [54, pp. 33–35] (Reprinted with permission of © Ilkka Piiroinen. All Rights Reserved)

Study, country, sample size	Participant characteristics	Follow-up years	Conclusions on SOC change
Blad et al. 2023, Sweden, $n = 415$	Age 28.2 (mean age at the childbirth). Mothers	25	Stable for the majority of women. Most stable in women with better health.
Sanna et al., 2022, Netherlands, $n = 489$	Age 70–97.	3	Stable even if experienced negative life events over lifespan.
Dziuba et al., 2021, Germany, $n = 392$	Age 33–67. Randomly selected residents.	20	Strengthened over 10 and 20 years of follow-up.
Wiesmann and Hannich, 2019, Germany, $n = 125$	Age 64–87. Active and healthy older individuals.	4	Strengthened, disclosing a small effect size.
Lindblad et al., 2016, Sweden, $n = 557$	Age 26–89. Female breast cancer patients.	3	Stable.
Silverstein and Heap, 2015, Sweden, $n = 1809$	Age 55–101. Combination of two panel studies.	10	Strengthened continuously into advanced old age. Decline after 70 if there are deficits in health and social resources.
Lövheim et al., 2013, Sweden, $n = 56$	Age > 85.	5	Strengthened. The more negative life events, the more weakening.
Bergman et al., 2011, Sweden, $n = 84$	Age 36–70. History of acute myocardial infarction.	2	Stable. Men and the low SOC group showed strengthening.
Feldt et al., 2011, Finland, $n = 18,525$	Age 20–54. Same cohort as in Feldt et al., 2007.	5	Strengthened in all age groups. High SOC, rather than higher age, determined a stable development.
Hochwälder and Forsell, 2011, Sweden, $n = 1012$	Age 30–64. Female nurses	1.5	Stable.
Liukkonen et al., 2010, Finland, $n = 11,521$	Age 18–62. Working age adults.	4–5	Unstable: The change was associated with type of employment trajectory.
Feldt et al., 2007, Finland, $n = 18,525$	Age 20–54. Working-age population.	5	Strengthened. More stable among subjects over 30 years than among younger adults. Strengthened more among younger than older age groups.
Hakanen, Feldt and Leskinen, 2007, Finland, $n = 532$	Age 44–57. Employees.	13	Strengthened. Most stable among high baseline SOC individuals.

(continued)

Table 4.1 (continued)

Study, country, sample size	Participant characteristics	Follow-up years	Conclusions on SOC change
Volanen et al., 2007, Finland, n = 17,271	Age 20–54. Same cohort as in Feldt et al., 2007.	5	Weakening after negative life events in both sexes. Initially strong SOC was not more stable than initially mediocre or weak SOC.
Kalimo et al., 2003, Finland, n = 174	Age: Working age adults.	10	Strengthened if no burnout. Weakened if serious burnout.
Kuuppelomäki and Utriainen, 2003, Finland, n = 81	Age 18–24. Students	3	Stable in 65%; strengthened in 32%; and weakened in 6%.
Smith, Breslin and Beaton, 2003, Canada, n = 6790	Age 18–64. Randomly selected individuals.	4	Unstable: 58% changed more than 10%.
Snekkevik et al., 2003, Norway, n = 26	Age 18–68. Subjects with severe multiple traumas.	3	Stable. Some subjects showed large variations.
Karlsson, Berglin and Larsson et al., 2000, Sweden, n = 111	Age 41–60. Coronary artery bypass surgery patients.	1	Stable in 58,7%; strengthened in 14,7%; and weakened in 26,6%.
Kivimäki et al., 2000, Finland, n = 577	Age 20–56. Municipal employees.	4	Stable in both sexes.

Notes: Participants include men and women (referred to in this paper as "sexes" if not stated otherwise)

New Developments of Salutogenic Questionnaires

There is a new trend in SAL research, that is, a new generation of questionnaires is developed using the salutogenic approach for the development of new scales with new items. The following are just a few examples. Some of them need to be further validated to show their usefulness and applicability.

First, an example is the Salutogenic Health Indicator Scale (SHIS) [55]. The scale consists of 12 indicators covering nine health-related dimensions: perceived stress, illness, energy, physical function, state of morale, psychosomatic function, expression of feelings, cognitive ability, social capacity and self-realization.

Second, The Work Experience Measurement Scale (WEMS) [56], consists of 32 statements about experiences of work and one's work situation that are divided into six different areas. The statements are developed based on theories that can be linked to, for example, sense of coherence, flow, regenerative work, demand-control-support and effort-reward. The content is activities-oriented and since the statements are positive, the result should be interpreted in a salutogenic perspective.

Third, another example of new scales is the Salutogenic Wellness Promotion Scale—short form (SWPS-SF) [57]. This is a 7-item scale, based on the salutogenic theory and measuring the presence of health-promoting factors such as engagement in health-promoting actions related to physical, intellectual, social, emotional, spiritual, vocational and environmental factors. The results suggest this short test can provide valid information without burdening the respondents. The authors recommend additional tests be conducted to validate the SWPS-SF.

Fourth, a self-report instrument for measuring the process of meaning-making among patients suffering from cancer, the Chinese Cancer Coherence Scale, is an instrument based on the salutogenic theory [58]. This instrument can be seen as more contextual and situational than the original SOC scale.

Fifth, The Childbearing SOC (CSOC), not to be confused with the child SOC (CSOC) is a scale measuring an individual's perception of the stresses, resources and meaningfulness of childbearing [59].

Finally, the Salutogenic Survey on Sustainable Working Life for Nurses (SalWork-N), explores and psychometrically tests a new profession-specific questionnaire identifying generalized and specific resistance resources against work-related stress among nurses [60]. The structure of the questionnaire indicates its usefulness in clinical practice for measuring resistance resources.

Towards a Revised Definition of the Sense of Coherence

More than 40 years have passed since Antonovsky (1) defined the SOC. It is perhaps time to revise the definition of SOC, based on the extensive research on the SOC concept after Antonovsky, here is an attempt, and further development is appreciated:

> a *multidimensional and culturally sensitive concept of* life orientation, *represented in an ease/dis-ease continuum*, that expresses the extent to which one has a pervasive, enduring though dynamic feeling of confidence that (1) the stimuli from one's internal and external environments in the course of living are structured, predictable, and explicable; (2) the resources are available to one to meet the demands posed by these stimuli *depending on the cultural context*; and (3) these demands are challenges, worthy of investment and engagement.

References

1. Antonovsky, A. (1987). *Unraveling the mystery of health. How people manage stress and stay well.* Jossey-Bass.
2. Lindström, B., & Eriksson, M. (2010). *The hitchhiker's guide to salutogenesis. Salutogenic pathways to health promotion.* The IUHPE Global Working Group on Salutogenesis and Folkhälsan Research Center.
3. Bergman, E., Malm, D., Ljungquist, B., Berterö, C., & Karlsson, J.-E. (2012). Meaningfulness is not the most important component for changes in sense of coherence. *European Journal of Cardiovascular Nursing, 11*(3), 331–338. https://doi.org/10.1016/j.ejcnurse.2011.05.005

4. Klepp, O. M., Mastekaasa, A., Sørensen, T., Sandanger, I., & Kleiner, R. (2007). Structure analysis of Antonovsky's sense of coherence from an epidemiological mental health survey with a brief nine-item sense of coherence scale. *International Journal of Methods in Psychiatric Research, 16*(1), 11–22. https://doi.org/10.1002/mpr.197
5. Drageset, J., & Haugan, G. (2016). Psychometric properties of the orientation to life questionnaire in nursing home residents. *Scandinavian Journal of Caring Sciences, 30*(3), 623–630. https://doi.org/10.1111/scs.12271
6. Rajesh, G., Eriksson, M., Pai, K., Seemanthini, S., Naik, D. G., & Rao, A. (2015). The validity and reliability of the sense of coherence scale among Indian university students. *Global Health Promotion, 23*, 16–26. https://doi.org/10.1177/1757975915572691
7. Söderhamn, U., Sundsli, K., Cliffordson, C., & Dale, B. (2015). Psychometric properties of Antonovsky's 29-item sense of coherence scale in research on older home-dwelling Norwegians. *Scandinavian Journal of Public Health, 43*(8), 867–874. https://doi.org/10.1177/1403494815598863
8. Eriksson, M., & Lindström, B. (2005). Validity of Antonovsky's sense of coherence scale: A systematic review. *Journal of Epidemiology & Community Health, 59*, 460–466. https://doi.org/10.1136/jech.2003.01808
9. Feldt, T., Leskinen, E., Koskenvuo, M., Suominen, S., Vahtera, J., et al. (2011). Development of sense of coherence in adulthood: A person-centered approach. The population-based HeSSup cohort study. *Quality of Life Research: An International Journal of Quality of Life Aspects of Treatment Care and Rehabilitation, 20*(1), 69–79. https://doi.org/10.1007/s11136-010-9720-7
10. Naaldenberg, J., Tobi, H., van den Esker, F., & Vaandrager, L. (2011). Psychometric properties of the OLQ-13 scale to measure sense of coherence in a community-dwelling older population. *Health and Quality of Life Outcomes, 23*(9), 37. https://doi.org/10.1186/1477-7525-9-37
11. Antonovsky, A. (1993). The structure and properties of the sense of coherence scale. *Social Science and Medicine, 36*(6), 725–733. https://doi.org/10.1016/0277-9536(93)90033
12. Schmalbach, B., Tibubos, A. N., Zenger, M., Hinz, A., & Brähler, E. (2020). Psychometric evaluation and norm values of an ultra-short version of the sense of coherence scale "SOC-3". *Psychotherapie Psychosomatik Medizinische Psychologie, 70*(2), 86–93. https://doi.org/10.1055/a-0901-7054
13. Bonaccorsi, G., Zanobini, P., Cosma, C., Buscemi, P., Paoli, S., Lastrucci, V., et al. (2023). Do demographic and socio-economic factors predict sense of coherence among university students? *Annali dell'Istituto Superiore di Sanità, 59*, 251–259.
14. Schumacher, J., Wilz, G., Gunzelmann, T., & Brähler, E. (2000). The Antonovsky sense of coherence scale. Test statistical evaluation of a representative population sample and construction of a brief scale. *Psychotherapie, Psychosomatik, Medizinische Psychologie, 50*(12), 472–482. https://doi.org/10.1055/s-2000-9207
15. Bargehr, B., Fischer von Weikersthal, L., Junghans, C., Zomorodbakhsch, B., Stolls, C., Prott, F.-J., et al. (2023). Sense of coherence and its context with demographics, psychological aspects, lifestyle, complementary and alternative medicine and lay aetiology. *Journal of Cancer Research and Clinical Oncology, 149*, 8393–8402. https://doi.org/10.1007/s00432-023-04760-9
16. Solfrank, M., Nikendei, C., Zehetmair, C., Friederich, H. M., & Nagy, E. (2023). The burden of substance use and (mental) distress among asylum seekers: A cross sectional study. *Frontiers in Psychiatry, 14*, 1258140. https://doi.org/10.3389/fpsyt.2023.1258140
17. Piiroinen, I., Tuomainen, T.-P., Tolmunen, T., & Voutilainen, A. (2024). Meaningfulness and mortality: Exploring the sense of coherence in Eastern Finnish men. *Scandinavian Journal of Public Health, 53*, 15. https://doi.org/10.1177/14034948231220091
18. Muroi, K., Ishitsuka, M., Hori, D., Doki, S., Ikeda, T., Takahashi, T., et al. (2023). A high sense of coherence can mitigate suicidal ideation associated with insomnia. *Health Psychology Report, 11*, 4. https://doi.org/10.5114/hpr/163068
19. Sebo, P., Rudrej, B., Bernard, A., Delaunay, B., Dupuy, A., Malavergne, C., et al. (2024). Validation and refinement of the sense of coherence scale for a french population: Observational study. *Interactive Journal of Medical Research, 13*, e50284. https://doi.org/10.2196/50284

20. Togari, T., Yamazaki, Y., Nakayama, K., & Shimizu, J. (2007). Development of a short version of the sense of coherence scale for population survey. *Journal of Epidemiology and Community Health, 61*, 921–922. https://doi.org/10.1136/jech.2006.056697
21. Skaug, E., Czajkowski, N. O., Waaktaar, T., & Torgersen, S. (2022). The role of sense of coherence and loneliness in borderline personality disorder traits: A longitudinal twin study. *Borderline Personality Disorder and Emotion Dysregulation, 9*, 19. https://doi.org/10.1186/s40479-022-00190-0
22. Margalit, M., & Efrati, M. (1996). Loneliness, coherence and companionship among children with learning disorder. *Educational Psychology, 16*(1), 69–80. https://doi.org/10.1080/0144341960160106
23. Idan, O., & Margalit, M. (2014). Socioemotional self-perceptions, family climate, and hopeful thinking among students with learning disabilities and typically achieving students from the same classes. *Journal of Learning Disabilities, 47*(2), 136–152. https://doi.org/10.1177/0022219412439608
24. Antonovsky, A., & Sourani, T. (1988). Family sense of coherence and family adaptation. *Journal of Marriage and the Family, 50*(1), 79–92. https://doi.org/10.2307/352429
25. Sagy, S., & Antonovsky, A. (1992). The family sense of coherence and the retirement transition. *Journal of Marriage and the Family, 54*, 983–993. https://doi.org/10.2307/353177
26. Sagy, S., & Antonovsky, H. (2000). The development of the sense of coherence: A retrospective study of early life experiences in the family. *Journal of Aging and Human Development, 51*(2), 155–166. https://doi.org/10.2190/765L-K6NV-JK52-UFKT
27. Vogt, K., Jenny, G. J., & Bauer, G. F. (2013). Comprehensibility, manageability and meaningfulness at work: Construct validity of a scale measuring work-related sense of coherence. *SA Journal of Industrial Psychology, 39*(1), 1–8. https://doi.org/10.4102/sajip.v39i1.1111
28. Hiemstra, S. R., Naaldenberg, J., De Jonge, A., & Vaandrager, L. (2024). Salutogenic mechanisms in nature-based work: Fostering sense of coherence for employees with limited capability for work. *Health Promotion International, 39*, daae127. https://doi.org/10.1093/heapro/daae127
29. Braun-Lewensohn, O., & Sagy, S. (2011). Salutogenesis and culture: Personal and community sense of coherence among adolescents belonging to three different cultural groups. *International Review of Psychiatry, 23*(6), 533–541. https://doi.org/10.3109/09540261.2011.637905
30. Eriksson, M., & Contu, P. (2022). The sense of coherence: Measurement issues. In M. B. Mittelmark, G. F. Bauer, L. Vaandrager, J. M. Pelikan, S. Sagy, M. Eriksson, et al. (Eds.), *The handbook of salutogenesis* (2nd ed., pp. 79–91). Springer.
31. Lanzara, R., Conti, C., Rosa, I., Pawłowski, T., Malecka, M., Rymaszewska, J., et al. (2023). Changes in hospital staff' mental health during the Covid-19 pandemic: Longitudinal results from the international COPE-CORONA study. *PLoS One, 18*(11), e0285296. https://doi.org/10.1371/journal.pone.0285296
32. Moons, P., Apers, S., Kovacs, A. H., Thomet, C., Budts, W., Enomoto, J., et al. (2021). Sense of coherence in adults with congenital heart disease in 15 countries. Patient characteristics, cultural dimensions and quality of life. *European Journal of Cardiovascular Nursing, 20*, 48–55. https://doi.org/10.1177/1474515120930496
33. Shorey, S., Ang, E., Baridwan, N. S., Bonito, S. R., Dones, L. B. P., Flores, J. L. A., et al. (2022). Salutogenesis and COVID-19 pandemic impacting nursing education across SEANERN affiliated universities: A multi-national study. *Nurse Education Today, 110*, 105277. https://doi.org/10.1016/j.nedt.2022.105277
34. Gómez-Salgado, J., Arias-Ulloa, C. A., Ortega-Moreno, M., García-Iglesias, J. J., Escobar-Segovia, K., & Ruiz-Frutos, C. (2022). Sense of coherence in healthcare workers during the covid19 pandemic in Ecuador: Association with work engagement, work environment and psychological distress factors. *International Journal of Public Health, 67*, 1605428. https://doi.org/10.3389/ijph.2022.1605428
35. Gomez-Salgado, J., Delgado-García, D., Ortega-Moreno, M., Fagundo-Rivera, J., El Khoury-Moreno, L., Vilches-Arenas, A., et al. (2024). Work engagement and sense of coherence as

predictors of psychological distress during the first phase of the COVID-19 pandemic in Chile. *Heliyon, 10*, e31327. https://doi.org/10.1016/j.heliyon.2024.e31327
36. Amoako, I., Srem-Sai, M., Quansah, F., Anin, S., Agormedah, E. K., Hagan, J. E., & Jnr. (2023). Moderation modelling of COVID-19 digital health literacy and sense of coherence across subjective social class and age among university students in Ghana. *BMC Psychology, 11*, 337. https://doi.org/10.1186/s40359-023-01334-9
37. Sawma, T., & Sanjab, Y. (2022). The association between sense of coherence and quality of life: A cross-sectional study in a sample of patients on haemodialysis. *BMC Psychology, 10*, 100. https://doi.org/10.1186/s40359-022-00805-9
38. Slootjes, J., Keuzenkamp, S. S., & Saharso, S. (2017). The mechanisms behind the formation of a strong sense of coherence (SOC): The role of migration and integration. *Scandinavian Journal of Psychology, 58*(6), 571–580. https://doi.org/10.1111/sjop.12400
39. Corless, I. B., Guarino, A. J., Nicholas, P. K., Tyer-Viola, L., Kirsey, K., Brion, J., et al. (2013). Mediators of antiretroviral adherence: A multisite international study. *AIDS Care, 25*(3), 364–377. https://doi.org/10.1080/09540121.2012.701723
40. Zakar, R., Iqbal, S., Zakar, M. Z., & Fischer, F. (2021). COVID-19 and health information seeking behavior: Digital health literacy survey amongst university students in Pakistan. *International Journal of Environmental Research and Public Health, 18*, 4009. https://doi.org/10.3390/ijerph18084009
41. Leung, A. Y. M., Parial, L. L., Tolabing, M. C., Sim, T., Mo, P., Okan, O., et al. (2022). Sense of coherence mediates the relationship between digital health literacy and anxiety about the future in aging population during the COVID-19 pandemic: A path analysis. *Aging & Mental Health, 26*(3), 544–553. https://doi.org/10.1080/13607863.2020.1870206
42. Alharbi, F. S., Aljemaiah, A. I., & Osman, M. (2022). Validation of the factor structure and psychometric characteristics of the Arabic adaptation of the sense of coherence SOC-13 scale: A confirmatory factor analysis BMC. *Psychology, 10*, 115. https://doi.org/10.1186/s40359-022-00826-4
43. Al-Yateem, N., Alrimawi, I., Fakhry, R., AlShujairi, A., Rahman, S. A., Marzougi, A. A., et al. (2021). Exploring the reliability and validity of the adapted arabic sense of coherence scale. *Journal of Nursing Measurement, 29*(2), E110–E125. https://doi.org/10.1891/JNM-D-19-00107
44. Antonovsky, A. (1993). The structure and properties of the sense of coherence scale. *Social Science & Medicine, 36*(6), 725–733. https://doi.org/10.1016/0277-9536(93)90033-z
45. Sirkiä, C., Laakkonen, E., Nordenswan, E., Karlsson, L., Korja, R., Karlsson, H., et al. (2024). Sense of coherence, its components and depressive and anxiety symptoms in expecting women and their partners—A FinnBrain birth cohort study. *Sexual & Reproductive Healthcare, 39*, 100930. https://doi.org/10.1016/j.srhc.2023.100930
46. Mafla, A. C., Herrera-López, M., España-Fuelagan, K., Ramírez-Solarte, I., Gallardo Pino, C., & Schwendicke, F. (2021). Psychometric properties of the SOC-13 scale in Colombian adults. *International Journal of Environmental Research and Public Health, 18*(24), 13017. https://doi.org/10.3390/ijerph182413017
47. Aune, I., Dahlberg, U., & Haugan, G. (2016). Sense of coherence among healthy Norwegian women in postnatal care: Dimensionality reliability and construct validity of the orientation to life questionnaire. *Sexual & Reproductive Healthcare, 8*, 6–12. https://doi.org/10.1016/j.srhc.2015.12.001
48. Carneiro, F. A. T., Salvador, V. F., Costa, P. A., & Leal, I. P. (2022). Family sense of coherence scale: A confirmatory factor analysis in a portuguese sample. *Frontiers in Psychology, 12*, 762357. https://doi.org/10.3389/fpsyg.2021.762357
49. Feldt, T., Lintula, H., Suominen, S., Koskenvuo, M., Vahtera, J., & Kivimäki, M. (2007). Structural validity and temporal stability of the 13-item sense of coherence scale: Prospective evidence from the population-based HeSSup study. *Quality of Life Research, 16*(3), 483–493. https://doi.org/10.1007/s11136-006-9130-z

50. Ding, Y., Bao, L.-P., Xu, H., Hu, Y., & Rahm-Hallberg, I. (2012). Psychometric properties of the Chinese version of sense of coherence scale in women with cervical cancer. *Psycho-Oncology, 21*(11), 1205–1214. https://doi.org/10.1002/pon.2029
51. Eriksson, M., & Lindström, B. (2006). Antonovsky's sense of coherence scale and the relation with health: A systematic review. *Journal of Epidemiology & Community Health, 60*(5), 376–381. https://doi.org/10.1136/jech.2005.041616
52. Hakanen, J. J., Feldt, T., & Leskinen, E. (2007). Change and stability of sense of coherence in adulthood: Longitudinal evidence from the healthy child study. *Journal of Research in Personality, 41*(3), 602–617. https://doi.org/10.1016/j.jrp.2006.07.001
53. Eriksson, M., & Lindström, B. (2007). Antonovsky's sense of coherence scale and its relation with quality of life—A systematic review. *Journal of Epidemiology and Community Health, 61*(11), 938–944. https://doi.org/10.1136/jech.2006.056028
54. Piiroinen, I. (2024). *Associations between sense of coherence and mortality: A study based on data from the Kuopio Ischaemic Heart Disease Risk Factor Study* (Vol. 838, p. 111). University of Eastern Finland Publications of the University of Eastern Finland Dissertations in Health Sciences.
55. Bringsén, Å., Andersson, H. I., & Ejlertsson, G. (2009). Development and quality analysis of the Salutogenic Health Indicator Scale (SHIS). *Scandinavian Journal of Public Health, 37*(1), 13–19. https://doi.org/10.1177/1403494808098
56. Nilsson, P., Andersson, H. I., & Ejlertsson, G. (2013). The work experience measurement scale (WEMS). *Work, 45*(3), 379–387. https://doi.org/10.3233/WOR-121541
57. Becker, C. M., Bian, H., Martin, R. J., Sewell, K., Stellefson, M., & Chaney, B. (2022). Development and field test of the Salutogenic Wellness Promotion Scale—Short form (SWPS-SF) in U.S. college students. *Global Health Promotion, 30*(1), 16–22. https://doi.org/10.1177/17579759221102
58. Chan, T. H. Y., Ho, R. T. H., & Chan, C. L. W. (2007). Developing an outcome measurement for meaning-making intervention with Chinese cancer patients. *Psycho-Oncology, 16*, 843–850. https://doi.org/10.1002/pon.1134
59. Li, B., Zhao, M., Zhu, Z., Zhao, H., Zhang, X., Wang, J., et al. (2024). The childbearing sense of coherence scale (CSOC-scale): Development and validation. *BMC Public Health, 24*, 1613. https://doi.org/10.1186/s12889-024-19109-1
60. Eriksson, M., Johannesson, E., Kerekes, N., Emilsson, M., Pennbrant, S., et al. (2024). Development and psychometric test of the salutogenic survey on sustainable working life for nurses: Identifying resistance resources against stress. *International. Journal Environmental Research and Public Health, 21*, 198. https://doi.org/10.3390/ijerph21020198

Open Access This chapter is licensed under the terms of the Creative Commons Attribution-NonCommercial-NoDerivatives 4.0 International License (http://creativecommons.org/licenses/by-nc-nd/4.0/), which permits any noncommercial use, sharing, distribution and reproduction in any medium or format, as long as you give appropriate credit to the original author(s) and the source, provide a link to the Creative Commons license and indicate if you modified the licensed material. You do not have permission under this license to share adapted material derived from this chapter or parts of it.

The images or other third party material in this chapter are included in the chapter's Creative Commons license, unless indicated otherwise in a credit line to the material. If material is not included in the chapter's Creative Commons license and your intended use is not permitted by statutory regulation or exceeds the permitted use, you will need to obtain permission directly from the copyright holder.

Chapter 5
Health, Mental Health and Quality of Life

Monica Eriksson and Eva Langeland

Health

The starting point for this chapter is to explore how health, mental health and quality of life (QoL) are defined. Health is a complex concept with different meanings depending on different perspectives on health. This chapter starts with defining what is meant by health, mental health and quality of life (QoL) based on the salutogenic theory and the contemporary evidence base of research after Antonovsky died [1, 2].

Health can either be considered as absolute, *"health is a state of complete physical, mental and social well-being and not merely the absence of disease infirmity."* [3] or as a resource-oriented concept, expressed in the Ottawa Charter for health promotion, *"health is a resource for everyday life, not the objective of living. It is a positive concept emphasizing social and personal resources, as well as physical capacities* [4]. Health is a precondition, an outcome and an indicator of sustainable societies [5].

The Global Sustainable Development Report [5] adopts a salutogenic approach to the health concept. This means that the concept of health integrates physical, mental, social and spiritual health on an individual (micro), group (meso) or societal (macro) level (see Chap. 1). It emphasizes the importance of structured and empowering environments, where people can identify their internal and external resources, use and reuse them to realize aspirations, to satisfy needs, to perceive meaningfulness, manage stress and to cope with changes in a health promoting manner. Based

M. Eriksson (✉)
Department of Psychology, Lund University, Lund, Sweden
e-mail: monica.eriksson@psy.lu.se

E. Langeland
Department of Health and Caring Sciences, Faculty of Health and Social Sciences, Western Norway University of Applied Sciences, Bergen, Norway

on the findings from a systematic review of salutogenesis and SOC, a certain possibility to modify and extend the health construct is becoming significant, implicating a health construct including salutogenesis and quality of life [6–9].

Life Promotion as an Expanded Definition of Health and Health Promotion

The idea is to expand the existing definition of health by integrating the principles of health promotion (the Ottawa Charter) and the most recent convention on human rights, i.e. the Convention of the Rights of the Child [10] with Antonovsky's salutogenic concepts. Health is an essential part of life satisfaction; however, quality of life and life satisfaction go beyond good health. Therefore, a modified definition of health promotion to *life promotion* towards well-being and sustainability is suggested, as follows:

> Life promotion is the process of enabling individuals, groups or societies to increase a feeling of confidence and thus maintain and improve their physical, mental, social and spiritual health. This could be reached by creating environments and societies characterized by clear structures and empowering environments where people see themselves as active participating individuals who are able to identify their internal and external resources, use and reuse them to realize aspirations, to satisfy needs, to perceive meaningfulness and to change or cope with the environment in a life promoting manner.

Mental Health

In the World Mental Health Report [11], the World Health Organization describes the state of mental health today. Mental health needs are high; however, services and health care are insufficient and inadequate. In most countries, poor mental health conditions are highly prevalent [11, p. 13]. There are three main reasons to invest in mental health: public health, human rights and socioeconomic development. Investing in mental health can enable social and economic development. Poor mental health slows down development by reducing productivity, straining social relationships and worsening cycles of poverty and disadvantage. Conversely, when people are mentally healthy and live in supportive environments, they can learn and work well and contribute to their communities, to the benefit of all [11, p. xvi].

Mental health is not about the presence or absence of mental disorder. Furthermore, mental health is not a binary state: we are not either mentally healthy or mentally ill. Thus, according to WHO mental health is defined as follows

> a state of mental well-being that enables people to cope with the stresses of life, realize their abilities, learn well and work well, and contribute to their community. Mental health is an integral component of health and well-being, is more than the absence of mental disorder.

It underpins our individual and collective abilities to make decisions, build relationships and shape the world we live in [11, p. 8].

This definition of mental health as a *state* of mental well-being differs from the salutogenic theory, which rejects health as a state. Instead, health is seen as a process on a continuum of ease/dis-ease. In a lecture held in Berkeley in 1993 Aaron Antonovsky conceptually defined salutogenesis as "*the continuing movement in an ease/dis-ease continuum*" [12]. The use of different resistance resources determines where we are in this continuum. Mental health refers to a person's position at any point in the life cycle on "*… a continuum that ranges from excruciating emotional pain and total psychological malfunctioning, at one extreme, to a full, vibrant sense of psychological well-being at the other*" [13, p. 274]. Further, Antonovsky described the movement on the continuum towards better mental health as shifting: "*From the use of unconscious psychological defence mechanisms toward the use of conscious coping mechanisms; from the rigidity of defensive structures to the capacity for constant and creative inner readjustment and growth; from a waste of emotional energy toward its productive use; from emotional suffering toward joy; from narcissism toward giving of oneself; and from exploitation of others to reciprocal interaction* [13, p. 274]".

In this publication a *two continua model* is used to describe and define mental health, that is as flourishing (a higher level of emotional, psychological and social well-being, that is good mental health) and languishing (people feel life meaningless, empty, etc., that is poor mental health) [14–17]. This means that people may have signs of mental health and symptoms of mental illness over a period. This implies the need for health professionals to support a reduction of the symptoms of mental illness and an improvement of the characteristics of mental health. Signs of mental health and symptoms of mental illness are two different dimensions that have a relationship which, together, constitute the overall level of mental health [14]. Mental health in terms of flourishing and languishing is shown in Fig. 5.1.

Interviewing persons with long-term mental disorders revealed experiences as a movement, like walking up and down a staircase. It was expressed both verbally in everyday language and through body language. Mental health for these persons is "*an aspect of being that is always present, and which is nourished by four domains of life: the emotional; physical; social and spiritual domains*" [16, p. 221]. A salutogenic perspective on successful rehabilitation after burnout was applied in a study aiming to examine what kind of resources could be considered as resistance resources against stress (GRRs, SRRs) and thereby supporting the rehabilitation process among young employees from The Netherlands [18]. In this qualitative study social support from family, friends and colleagues, as well as having a feeling of control over the rehabilitation process, seemed to be the key resources in facilitating a stable, meaningful return to work after burnout.

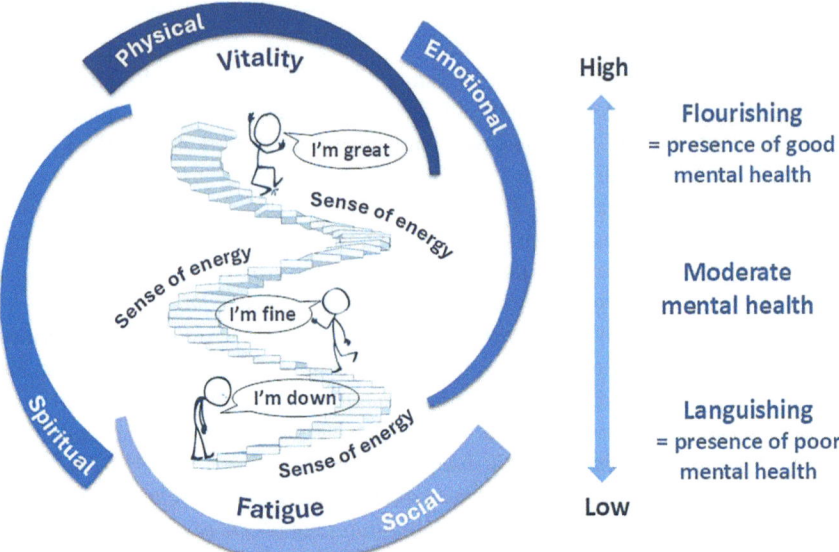

Fig. 5.1 Perceived mental health—a sense of energy in a continuum ranging from flourishing (good mental health) to languishing (poor mental health) [17]. (Reprinted with permission of © Nina Helen Mjøsund. All Rights Reserved)

Quality of Life

Quality of life (QoL) has been defined as personal well-being or satisfaction with life [19], as well as physical and material well-being, relations with other people, social, communal, civic activities, personal development and fulfilment, positive mental health [20] and as related to health (HRQOL) [21]. Functional status, often actually functional limitations, and health are two dimensions of HRQOL. Both QoL and health are complex concepts. The definition of QoL by the WHO QoL Assessment Group captures physical and mental health in terms of positive development aspects of health like coping, resilience, satisfaction and autonomy [20]. In the WHO Health Promotion Glossary Nutbeam [22, p. 361] defines QoL as *"the perception of individuals that their needs are being satisfied and that they are not being denied opportunities to achieve happiness and fulfilment, regardless of physical health status, or social and economic conditions."*

A salutogenic interpretation of the QoL concept combines the global, external, interpersonal and personal resources of an individual, group or society [23, 24]. These four dimensions form a holistic definition of QoL based on the salutogenic theory as follows:

> QoL is the total existence of an individual, a group or a society describing the essence of existence as measured objectively and perceived subjectively by the individual, group or society. [24, p. 43].

Having the definitions of health, mental health and QoL in place it is time to explore the role of SOC in relation to health and QoL. Research on these concepts is extensive, therefore, trying to get a reasonable overview and for the purposes of this brief publication, the focus has mainly been on scientific review papers published after 2003 and until 2024.

A Population Perspective: The Current State of Knowledge

A strong SOC is associated with good health, particularly mental health, through its mediating and moderating effect on perceived stress. A strong SOC seems to protect against anxiety, depression, burnout and hopelessness and is strongly and positively related to health resources such as optimism, hardiness, control and coping. Further, a strong SOC predicts good health and QoL from childhood to adulthood [7, 8]. Exploring how SOC develops during adolescence and how it is linked to health, mental health and psycho-social behaviour were the objective of an extensive review of studies around the world in the last decade [25, p. 134]. This study showed that personal and systemic (i.e. family and community) SOC are meaningful resources for effective coping with a wide variety of stressful situations in different cultures. However, there are other results reported, where SOC among university students did not buffer but mediated the effects of stressors on wellbeing [26, p. 203]. Students especially suffered from reduced feelings of manageability when confronted with financial strains, dissatisfying study situations, or disrupted plans.

Research on large-scale population studies on health is limited. The focus has been on risk factors for various diseases and an overall mortality rate, such as in The Kuopio Ischaemic Heart Disease Risk Factor Study *amongst Finnish adult men (n = 48,138, median follow-up = 14.1 years). SOC has here been used to measure the relationship between SOC and mortality.* Piiroinen [27] explored the role of SOC and the dynamics of the SOC components (comprehensibility, manageability and meaningfulness) and their associations with mortality. The findings underscore the role of SOC in health. A weak SOC was here linked to higher mortality risks than in the general adult population. Furthermore, the findings suggest that maintaining or stabilizing SOC could enhance health and longevity. Among SOC components, meaningfulness emerges as a key predictor of mortality. Together these findings support Antonovsky's theory that a strong SOC, and especially meaningfulness, can act as protective against life's stress and promote health.

However, there are also opposite findings. By face-to-face interviews with 464 Bedouin women in Israel depressive symptoms (DS) and two levels of SOC (low/high) were explored [28]. The aim was to examine the contribution of SOC to predict DS beyond psychological resources and socioeconomic position. While strong SOC was positively and significantly associated with DS, SOC was not associated with DS in the weak SOC group. Thus, the relationships between strong versus weak SOC and DS among Bedouin women differ from those found in Western societies. This is once again evidence for questioning the use of SOC as a universal tool in different cultural contexts (see Chaps. 4 and 11).

In a Norwegian sample of home-living residents with mental health problems the predictive value of SOC was assessed at baseline and a 1-year follow up. SOC seemed here to predict changes in life satisfaction, but mental health problems did not [29]. These findings emphasize the importance of assessing factors that may explain differences in life satisfaction over and above mental symptoms among people with MHP. The results indicate that improving SOC among people with MHP might provide important opportunities for improving their life satisfaction.

A Professional Perspective: The Current State of Knowledge

Extensive research describes how nurses experience their work environment as stressful and dissatisfying with nurses intending to leave their workplace. Turnover among nurses is a global concern that negatively affects health services. In the European context, there is a serious nursing shortage in most countries [30]. The "Registered nurse forecasting in Europe study" (RN4CAST) [31] brought together researchers from 12 European countries to generate a large evidence basis about nursing workforce and nurses' intention to leave the workplace [32, 33]. The study aimed to explore the factors that increase both the professional longevity and nurses' willingness to remain in work and the profession. Therefore, a qualitative study on nurses in western Sweden ($n = 12$) was conducted, aiming to explore and explain factors that were important for nurses to remain in the workplace [34]. The results showed that within the three themes of coherence (comprehensibility, manageability and meaningfulness), job satisfaction and fun at work, manageable workload, collaboration and supportive leadership among others explained why nurses remained in the workplace.

In a systematic review about the relationship between SOC and work stress and well-being perceived by care professionals, González-Siles and colleagues [35] found that stress, depression, burnout and posttraumatic stress disorder (PTSD) negatively correlated with SOC; in contrast, job satisfaction, well-being and quality of life positively correlated with SOC. It was concluded that SOC could act as a mediating variable and as a predictor variable of these mental health problems. Pachi et al. [36] investigated the prevalence of depression and burnout and possible association with SOC amongst nursing staff during the COVID pandemic. Female nurses had higher burnout and depression scores compared to men and weaker levels of SOC. Mediation analysis indicated a partial mediation of burnout in the correlation between SOC and depression. The SOC acted as a negative moderator between burnout and depression, that is, the stronger the SOC the lower the level of burnout and depression.

SOC and Health Behaviours

The association between a strong SOC and healthy behaviours is interesting. Some health promotion professionals are much concerned about individual behaviour like alcohol use, drug abuse, eating behaviour and healthy food selection, physical exercise and oral health behaviour. Many activities in health promotion practice have been implemented to monitor individual behavioural change, underestimating the difficulty to achieve real long-lasting change of behaviour. Antonovsky did not use the concept healthy behaviour but used a related concept, "a healthy orientation," that served as a GRR (see Chap. 3). Combined with other GRRs a healthy orientation serves as a prerequisite for the development of a strong SOC [2].

The evidence demonstrates that SOC has an impact on healthy behaviours, the stronger the SOC the healthier behaviour. Results from a systematic review on the associations between SOC and health-related behaviours among adolescents and young adults in Spain identified eight health-related behaviours; alcohol use, physical activity, tobacco use, eating habits, rest periods, use of illegal substances, behaviours related to oral health and time spent in games on the computer [37]. The results support a strong relationship of SOC with healthy behaviours both as a protective factor against risk behaviours and for its positive association with preventive and health-promoting behaviours of adolescents, young adults and university students.

Independent of social class and the level of education individual differences in health behaviour and its relation to SOC are found [38]. This result is based on the EPIC-Norfolk study on the adult population in UK ($n = 18,287$). Persons with the strongest SOC were 28 percent less likely to smoke (OR 0.72), 36 percent less likely to be physically inactive (OR 0.64), and consumed on average more fruit and vegetables and fibre per day than those with the weakest SOC.

Concluding Remarks

To sum up, SOC seems to have an impact on health, mental health, QoL and health behaviour; the stronger the SOC, the better health and QoL and healthier behaviour. Furthermore, longitudinal studies confirm the predictive validity of the SOC for a good health development and QoL.

References

1. Antonovsky, A. (1979). *Health, stress and coping*. Jossey-Bass.
2. Antonovsky, A. (1987). *Unraveling the mystery of health. How people manage stress and stay well*. Jossey-Bass.
3. WHO. (1948). *Constitution*.

4. WHO. (1986). *Ottawa charter for health promotion: An international conference on health promotion, the move towards a new public health, November 17–21, 1986*. World Health Organization.
5. United Nations. (2023). *The global sustainable development report*. Retrieved October 12, 2024, from https://sdgs.un.org/goals
6. Eriksson, M., & Lindström, B. (2005). Validity of Antonovsky's sense of coherence scale—Systematic review. *Journal of Epidemiology and Community Health, 59*(6), 460–466. https://doi.org/10.1136/jech.2003.018085
7. Eriksson, M., & Lindström, B. (2006). Antonovsky's sense of coherence scale and the relation with health - A systematic review. *Journal of Epidemiology and Community Health, 60*, 376–381. https://doi.org/10.1136/jech.2005.041616
8. Eriksson, M., & Lindström, B. (2007). Antonovsky's sense of coherence scale and its relation with quality of life: A systematic review. *Journal of Epidemiology and Community Health, 61*(11), 938–944. https://doi.org/10.1136/jech.2006.056028
9. Lindström, B., & Eriksson, M. (2006). Contextualising salutogenesis and Antonovsky in public health. *Health Promotion International, 21*(3), 238–244. https://doi.org/10.1093/heapro/dal016
10. United Nations, UNCRC. (1989). *The United Nations convention on the rights of the child*. Retrieved October 11, 2024, from https://www.ohchr.org/sites/default/files/crc.pdf
11. WHO. (2022). *World mental health report, transforming mental health for all*. World Health Organization. Retrieved October 12, 2024, from https://iris.who.int/bitstream/handle/10665/356119/9789240049338-eng.pdf?sequence=1
12. Antonovsky, A. (1993). *The salutogenic approach to aging*. Lecture held in Berkeley, January 21, 1993. Retrieved Oktober 19, 2024, from https://www.angelfire.com/ok/soc/a-berkeley.html
13. Antonovsky, A. (1985). The life cycle, mental health and the sense of coherence. *Israel Journal of Psychiatry and Related Sciences, 22*(4), 273–280.
14. Keyes, C. L. M. (2005). Mental illness and/or mental health? Investigating axioms of the complete state model of health. *Journal of Consulting and Clinical Psychology, 73*(3), 539–548. https://doi.org/10.1037/0022-006X.73.3.539
15. Langeland, E., & Vinje, H. F. (2013). The significance of salutogenesis and well-being in mental health promotion: From theory to practice. In C. L. M. Keyes (Ed.), *Mental well-being. International contributions to the study of positive mental health* (pp. 299–329). Springer.
16. Mjøsund, N. H., Eriksson, M., Norheim, I., Keyes, C. L. M., Espnes, G. A., & Vinje, H. F. (2015). Mental health as perceived by persons with mental disorders—An interpretative phenomenological analysis study. *International Journal of Mental Health Promotion, 17*(4), 215–233. https://doi.org/10.1080/14623730.2015.1039329
17. Mjøsund, N. H. (2017). *Positive mental health - From what to how. A study in the specialized mental healthcare service*. Norwegian University of Science and Technology, Faculty of Medicine and Health Sciences, Department of Public Health and Nursing. https://ntnuopen.ntnu.no/ntnu-xmlui/handle/11250/2447606?locale-attribute=en
18. Pijpker, R., Vaandrager, L., Veen, E. J., & Koelen, M. (2021). Seizing and realizing the opportunity: A salutogenic perspective on rehabilitation after burnout. *Work, 68*(3), 551–561. https://doi.org/10.3233/WOR-203393
19. Fayers, P. M., & Machin, D. (2000). *Quality of life. Assessment, analysis and interpretation*. John Wiley & Sons.
20. Kovess-Masfety, V., Murray, M., & Gureje, O. (2005). Evolution of our understanding of positive mental health. In H. Herrman, S. Saxena, & R. Moodie (Eds.), *Promoting mental health. Concepts, emerging evidence, practice* (pp. 35–45). World Health Organization.
21. Kaplan, R. M., & Hays, R. D. (2022). Health-related quality of life measurement in public health. *Annual Review of Public Health, 43*, 355–373. https://doi.org/10.1146/annurev-publhealth-052120-012811
22. Nutbeam, D. (1998). Health promotion glossary. *Health Promotion International, 13*(4), 349–364. https://doi.org/10.1093/heapro/13.4.349

23. Lindström, B. (1992). Quality of life: A model for evaluating health for all. Conceptual considerations and policy implications. *Sozial und Praventivmedicin, 37*, 301–306. https://doi.org/10.1007/BF01299136
24. Lindström, B. (1994). *The essence of existence. On the quality of life of children in the Nordic countries—Theory and practice in public health*. Doctoral thesis, Nordic School of Public Health.
25. Braun-Lewensohn, O., Idan, O., Lindström, B., & Margalit, M. (2017). Salutogenesis: Sense of coherence in adolescence. In M. B. Mittelmark, S. Sagy, M. Eriksson, G. F. Bauer, J. M. Pelikan, B. Lindström, et al. (Eds.), *The handbook of salutogenesis* (2nd ed., pp. 123–136). Springer.
26. Kulcar, V., Kreh, A., Juen, B., & Siller, H. (2023). The role of sense of coherence during the COVID-19 crisis: Does it exercise a moderating or a mediating effect on university students' wellbeing? *SAGE Open, 13*(1), 21582440231160123. https://doi.org/10.1177/21582440231160123
27. Piiroinen, I. (2024). *Associations between sense of coherence and mortality: A study based on data from the Kuopio Ischaemic Heart Disease Risk Factor Study* (p. 838). University of Eastern Finland Publications of the University of Eastern Finland Dissertations in Health Sciences.
28. Daoud, N., Braun-Lewensohn, O., Eriksson, M., & Sagy, S. (2014). Sense of coherence and depressive symptoms among low-income Bedouin women in the Negev Israel. *Journal of Mental Health, 23*(6), 307–311. https://doi.org/10.3109/09638237.2014.951475
29. Langeland, E., Wahl, A. K., Kristoffersen, K., Nortvedt, M. W., & Hanestad, B. R. (2007). Sense of coherence predicts change in life satisfaction among home-living residents in the community with mental health problems: A one-year follow-up study. *Quality of Life Research, 16*(6), 939–946. https://doi.org/10.1007/s11136-007-9199-z
30. Zander, B., Aiken, L. H., Busse, R., Rafferty, A. M., Sermeus, W., & Bruyneel, L. (2016). Nursing in the European Union. *EuroHealth, 22*, 3–6.
31. RN4CAST. Retrieved October 22, 2024, from http://www.rn4cast.eu
32. Nowrouzi-Kia, O. T., & Fox, T. M. (2020). Factors associated with intent to leave in registered nurses working in acute care hospitals. A cross-sectional study in Ontario, Canada. *Workplace Health and Safety, 68*(3), 121–128. https://doi.org/10.1177/2165079919884956
33. Leineweber, C., Chungkham, H. S., Lindqvist, R., Westerlund, H., Runesdotter, S., Smeds Alenius, L., et al. (2016). Nurses' practice environment and satisfaction with schedule flexibility is related to intention to leave due to dissatisfaction: A multi-country, multi-level study. *International Journal of Nursing Studies, 58*, 47–59. https://doi.org/10.1016/j.ijnurstu.2016.02.003
34. Nunstedt, H., Eriksson, M., Obeid, A., Hillström, L., Truong, A., & Pennbrant, S. (2020). Salutary factors and hospital work environments: A qualitative study of nurses in Sweden. *BMC Nursing, 19*(1), 125. https://doi.org/10.1186/s12912-020-00521-y
35. González-Siles, P., Martí-Vilar, M., González-Sala, F., Merino-Soto, C., & Toledano-Toledano, F. (2022). Sense of coherence and work stress or well-being in care professionals: A systematic review. *Healthcare, 10*, 1347. https://doi.org/10.3390/healthcare10071347
36. Pachi, A., Sikaras, C., Ilias, I., Panagiotou, A., Zyga, S., Tsironi, M., et al. (2022). Burnout, depression and sense of coherence in nurses during the pandemic crisis. *Healthcare, 10*, 134. https://doi.org/10.3390/healthcare10010134
37. da Silva-Domingues, H., del Pino-Casado, R., Palomino-Moral, P. A., Martinez, C. L., Moreno-Camara, S., & Frias-Osuna, A. (2022). Relationship between sense of coherence and health-related behaviours in adolescents and young adults: A systematic review. *BMC Public Health, 22*(1), 477. https://doi.org/10.1186/s12889-022-12816-7
38. Wainwright, N. W. J., Surtees, P. G., Welch, A. A., Luben, R. N., Khaw, K.-T., & Bingham, S. A. (2007). Healthy lifestyle choices: Could sense of coherence aid health promotion? *Journal of Epidemiological and Community Health, 61*, 871–876. https://doi.org/10.1136/jech.2006.056275

Open Access This chapter is licensed under the terms of the Creative Commons Attribution-NonCommercial-NoDerivatives 4.0 International License (http://creativecommons.org/licenses/by-nc-nd/4.0/), which permits any noncommercial use, sharing, distribution and reproduction in any medium or format, as long as you give appropriate credit to the original author(s) and the source, provide a link to the Creative Commons license and indicate if you modified the licensed material. You do not have permission under this license to share adapted material derived from this chapter or parts of it.

The images or other third party material in this chapter are included in the chapter's Creative Commons license, unless indicated otherwise in a credit line to the material. If material is not included in the chapter's Creative Commons license and your intended use is not permitted by statutory regulation or exceeds the permitted use, you will need to obtain permission directly from the copyright holder.

Part II
Applications

Chapter 6
Salutogenesis in the Context of Health Care

Monica Eriksson and Eva Langeland

A Change of Perspective: Another Way of Thinking

Having the historical and the theoretical foundation of the salutogenic theory and the core concept sense of coherence (SOC) in place (see Chaps. 1–4), this chapter explains how the theory can guide practice in the context of health care. The starting point is once again a change of perspective, based on the guiding principles of salutogenesis and health promotion [1–3]. The salutogenic guiding principles constitute a way of thinking in terms of people's resources, a way of working, as well as a way of approaching and caring for other people [4]. This leads to a reorientation of the health care sector towards a resource-oriented thinking and acting, where the focus is on both professional's and the patient's health and resistance resources, as well as a capacity building in healthcare professionals [5, 6]. Vinje [7] introduced the concept of "self-tuning," referring to habitual self-sensitivity, reflection, and mobilizing of resources, which can play a central role in nursing care professionals who should strive to "live the talk," develop their personal salutogenic capacity—in other words, *do* what you say and *be* what you are as a salutogenic professional [1, 7], that is learning by being. The guiding principles and criteria for salutogenic interventions in the context of health care are shown in Table 6.1 [2, 3, 8, 9].

M. Eriksson (✉)
Department of Psychology, Lund University, Lund, Sweden
e-mail: monica.eriksson@psy.lu.se

E. Langeland
Department of Health and Caring Sciences, Faculty of Health and Social Sciences, Western Norway University of Applied Sciences, Bergen, Norway

Table 6.1 Guiding principles for salutogenic interventions in the context of health care

Guiding principles/salutogenic criteria	Focus	References
A focus on health-promoting factors: identifying generalized resistance resources (GRRs) and/or specific resistance resources (SRRs) and use and reuse these in a health promoting way.	Resistance resources	Langeland et al. [1]
A whole-person approach (WPA) which includes the life course within the life circumstances of the participant, the unique life story.	Holistic approach	
Active adaptation (A) includes the target participant's ability to actively participate in the interplay between the person/group and surrounding environment.	Become actively involved	
Stressors and tension as normal experiences of life and as potentially health-promoting (ST), which transform stressors and tension into coping.	Appropriate challenges	
SOC as a learning process (L), where the resistance resources are identified and a learned capability to use them, which makes learning experiences turn into coping.	Learning	
Developing not only a sense *of* coherence but a sense *for* coherence, that is, an ability to understand and know *how* to strengthen SOC and its effects.	SFC/SOC	Koelen & Lindström [10] Magistretti et al. [11]
A synergy approach between interventions and activities by supporting decision-focused, multi-professional, and multi-disciplinary work.	Synergy	Péres-Wilson et al. [12]
Adopt a coherent system approach characterized by sustainability and an environmental ecological thinking.	Sustainability	Lindström & Eriksson [13]
Empowerment as the key mechanism for the health promoting processes.	Empowerment	
Active participation in interventions that affect the individual.	Participatory approach	
Ethical values based on Human Rights and Convention of the Rights for the Child.	Ethics	

Salutogenic Interventions

In the middle of the 1990s, the first studies on how SOC can be strengthened by interventions presented [14, 15] and followed by several studies in the 2000s [16–18]. Langeland and colleagues [1, 8, 19, 20] conducted the first study aimed at evaluating the treatment based on salutogenic principles on coping (defined as a sense of coherence) of people with mental health problems, using a RTC study design. This study presented the details of an intervention program [8] on *how* to strengthen SOC, describing *how* to structure and implement the intervention program as well as explaining its basic ideas and theories. One of the key mechanisms for strengthening SOC is to let people learn to be more conscious of their strengths and resources, to improve their ability to use them, thus, shift towards a positive interplay between the use of internal and/or external resistance resources and SOC. This means there are many ways to promote SOC dependent on a person's or a group's needs and available resources [1, 3, 8].

There are different ways to classify salutogenic interventions. A scoping review [1, 2] showed that out of 41 intervention studies with SOC as an outcome only three studies met the requirement for all five criteria suggested by Langeland et al. [2] for salutogenic interventions [19, 21, 22]. Further, in an integrative review Guo and colleagues [23] synthesized findings from 18 eligible studies on older adults and found that interventions based on salutogenesis fell into three main categories: dialogue-based, health education courses based, goal setting and achievement based. The length of the intervention ranged from 4 weeks to 2 years, with most ($n = 12$) within 12 weeks; the duration of each session ranged from 30 to 150 min, with the majority ($n = 7$) within 1 h; the frequency ranged from five times weekly to three times in 10 months, and in six studies was once a week. Intervention providers were mostly multidisciplinary teams, while in four studies were nurses only. Most of the studies reported that salutogenic-based interventions improved older adults' sense of coherence, quality of life, self-efficacy, self-management, meaning of life and mental health.

In Table 6.2 a selection from the 41 studies included in the scoping review [2], different types of interventions show examples of practical applications demonstrating the diversity. Due to this variation the studies are not possible to compare. However, all studies reveal significant improvements in SOC due to active participation in programs [19, 21, 24–26].

Research indicates that it is possible to increase the SOC by participating in different kinds of interventions in various settings and of various lengths [2]. However, it is reasonable to think that by making a stronger salutogenic content, following the guiding principles in Table 6.1 in the interventions you thereby match both content and outcome better. The consequence is SOC becomes more sensitive to change and thus stronger. To sum up, we need more longitudinal salutogenic interventions studies with larger sample sizes and stronger research designs such as RCTs to increase the knowledge about SOC's other salutogenic outcomes' ability to change both in the short and long run. A general conclusion based on the presented interventions is that SOC can be strengthened and salutogenic outcomes can be learned. This gives hope for the future.

Implementation in Practice—"The Toolbox"

During years of lecturing on salutogenesis, one question constantly reoccurs, that is, where can the "toolbox" for implementation be found. The answer has always been the same, there is no universal "toolbox," and to be more precise, there will never be any *universal* manual for the application. To be able to apply the salutogenic theory in practice, it is necessary to *understand* the philosophical foundation of

Table 6.2 Overview of selected salutogenic intervention studies (Adapted from Langeland et al., 2022 [2], Table 20.1. Some modifications were made. https://doi.org/10.1007/978-3-030-79515-3_20, licensed under the terms of the Creative Commons Attribution 4.0 International License (http://creativecommons.org/licenses/by/4.0/))

First Author/Year/Country/Sample/Title	Methods Assessment points	Type and content of intervention(s)	Salutogenic criteria[a]	SOC scale	Results SOC[b]
Fagermoen, et al. 2015, Norway Patients, women and men, with morbid obesity ($N = 68$), mean age 43.4, (SD 10.3). *Morbid obese adults increased their sense of coherence 1 year after a patient education course: a longitudinal study* [24].	Follow up 12 months	A *patient education* course over a period over 9–12 weeks. It consisted of 40 h. Grounded in cognitive behaviour theory and emphasises the participants work in uncovering hidden resources and strengthening self-concept. It includes life-style changes, individual action plan, guided reflections, self-help group and physical activity	A, L, GRR, SRR, WPA	SOC-13 Primary outcome	SOC mean: 54.7 → 59.2 ($p = 0.0.001$) points
Höjdahl, 2015, Russia, Sweden, Norway, Estonia, Denmark Women in correctional institutions ($N = 534$) Age Median:34 *Emotional distress and sense of coherence in women completing a motivational program in five countries. A prospective study* [22].	Follow up 12 months	The *"VINN" program*: 15 3 h sessions. In groups combined with homework, they were encouraged to identify something meaningful that they can engage with in their personal lives while serving and after serving their sentences. Their personal motivation for and commitment to change behaviour was stimulated, within an atmosphere of acceptance and compassion.	A, L, SRR, ST, WPA	SOC-13 Primary outcome	SOC mean: 52.4 → 54.24. (95 percent CI (0.72, 2.92)

Langeland, et al. 2006, Norway Residents in the community with mental health challenges. Intervention (n = 42). Mean age:51(range:18–80) Control (n = 56). *The effect of salutogenic treatment principles on coping with mental health problems—A randomised controlled trial* [19].	RCT 6 and 12 months	*Salutogenic talk-therapy* groups once a week 16 times, each 2 h. In addition, homework. The main objective of the group intervention was to increase participants' consciousness of their potential, their internal and external resistance resources and their ability to use them, and thus to increase their coping in the context of everyday living. The program was adapted to the person's ability to work with her/himself and their relations. In addition, TAU. Control group: TAU	A, L, GRR, SRR, ST, WPA	SOC-29 Primary outcome	6 months: Intervention group SOC mean 120 → 125. Control group: 111 → 110. (P = 0.03, E.S = 0.29). After 12 months: P = 0.48
Malm et al., 2018, Sweden Older people with fibrillations. Intervention (n = 56). Control (n = 55): *Effects of brief mindfulness-based cognitive behavioural therapy on health-related quality of life and sense of coherence in atrial fibrillation patients* [25].	RCT 9 weeks and 12 months	*Brief dyadic cognitive behavioural therapy* (CBT) programme and TAU. CBT consisted of three 2.5-h group sessions over a period of 9 weeks stress reduction programme. It comprised a series of self-regulation techniques that facilitate, through the regulation of emotions and behaviours and psychological coherence and mindfulness practice. Control group: TAU	SRR, A, L	SOC-13 Secondary outcome	12 months: Intervention: SOC: 73.31 → 76.15 control: 70.24 → 70.08 (p = 0.04)
Merakou et al., 2018, Greece. People with long-term unemployment suffering from anxiety disorders. Mean age 33.5 (SD 6.4) Intervention group (n = 30). Control group (n = 20). *The effect of progressive muscle relaxation on emotional competence, depression, anxiety, stress, sense of coherence, health-related quality of life, and well-being of unemployed people in Greece: An intervention study*.	Follow-up with control group 8 weeks	The intervention consisted of an 8-week on *Progressive Muscle Relaxation* (PMR) counselling services. The participants of the intervention group were divided in four subgroups of 6–8 people with each small group attending a two-month training course for the PMR technique. The training included 4 weekly sessions (45 min) from a professional PMR trainer that took place during the first month. Control group: Counselling services.	A, L, GRR, SRR	SOC-13 Primary outcome	Intervention: 45.80 → 49.34 (p = 0.001) control: 53.68 → 54.65 (p = 0.008)

[a] *A* Active adaptation, *GRR* General Resistance Resources, *SRR* Specific Resistance Resources, *L* Learning, *ST* Tension and stressors as potentially health promoting, *WPA* Whole Person Approach
[b] *ES* effect size, *CI* Confidence Interval, *p* p-value, *TAU* Treatment As Usual

salutogenesis, and even more, internalize this new knowledge within yourself, only thereafter you are ready to *be*, *think* and *work* in a salutogenic way [1]. This process takes time. In the next the health-promoting salutogenic dialogue is described and explained, as a tool for health promotion the salutogenic way.

Dialogue as the Tool for Health Promotion

The salutogenic dialogue as a method and tool for health promotion a salutogenic way is through "The Collegial Model," previously developed for research supervision [27], here adapted for health care, and shown in Fig. 6.1. The guiding principles for salutogenic interventions in line with Table 6.1, are integrated in the figure.

This model is based on the core principles of the Ottawa Charter for Health Promotion, that is the Human Rights, and considering people as active participating subjects, enabling them to take control over their health determinants [28]. These principles are combined with theoretical and conceptual knowledge of salutogenic theory, as the core concepts Sense of Coherence, Sense FOR Coherence and the Generalized and Specific Resistance Resources [29–31]. The mode of thinking behind this model is to learn in a reciprocal learning process.

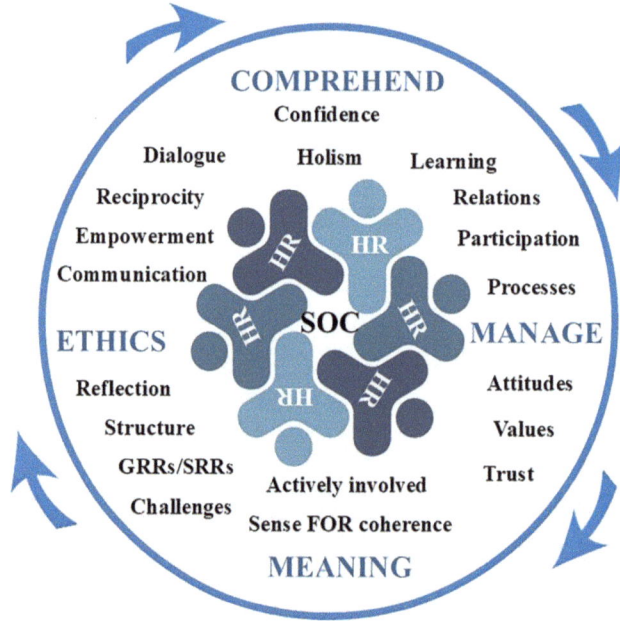

Fig. 6.1 "The Collegial Model" for salutogenic dialogues in the context of health care [27] (© The Author(s), 2018. Reproduced by permission of Oxford University Press. All Rights Reserved)

Placed in the middle of the circle are patients, participants for example in interventions, health professionals and relatives. In the Collegial Model, all parties involved are seen as colleagues or collaborators, each one has specific knowledge and different perspective, a holistic or whole-person perspective. They all form a collaborative team; they have different kinds of knowledge and experiences; all are given equal value. The participants need to see each other as equal partners in a collaborative learning. All involved share a common goal of maintaining and promoting health. The salutogenic dialogue is about relations in a certain context, here the healthcare sector, characterized by comprehensibility, manageability and meaningfulness. This relationship must be characterized by mutual trust, confidence and respect among all parties. A lack of confidence may destroy the entire process. Much has been written about the imbalance of power in relations between health professionals and participants. However, this model assumes that power is an unlimited entity, by giving power and responsibility to patients and relatives while still maintaining the power of the health professionals. Health professionals need to know about the patients' experiences and, patients need to be provided sufficient knowledge by the health professionals.

An example of targeted health dialogues in primary health care can be found in Sweden, where these dialogues have been implemented for several years [32]. The health dialogues are undertaken through a pedagogical approach allowing individuals to reflect on their resources, the situation and motivation to change their mode of lifestyle supported by healthcare professionals. Such dialogues have been systematically evaluated. Health professionals' experiences were explored through focus group interviews [32]. The results showed that health professionals considered the dialogues as a valuable opportunity to promote health in communities. The pedagogical tool including the visual health profile was experienced to have an important impact on the dialogue offering direction for actions to promote health.

References

1. Langeland, E., Hanson Ausland, L., Gunnarsdottir, H., Arveklev, S. H., & Vinje, H. F. (2022a). Promoting salutogenic capacity in health professionals. In M. B. Mittelmark, G. F. Bauer, L. Vaandrager, J. M. Pelikan, S. Sagy, M. Eriksson, et al. (Eds.), *The handbook of salutogenesis* (2nd ed., pp. 611–624). Springer.
2. Langeland, E., Vaandrager, L., Nilsen, A. B. V., Schraner, M., & Meier Magistretti, C. (2022b). Effectiveness of interventions to enhance the sense of coherence in the life course. In M. B. Mittelmark, G. F. Bauer, L. Vaandrager, J. M. Pelikan, S. Sagy, M. Eriksson, et al. (Eds.), *The handbook of salutogenesis* (2nd ed., pp. 201–219). Springer.
3. Langeland, E., Vinje, H. F. (2013). The significance of salutogenesis and well-being in mental health promotion: From theory to practice. In Keyes C., Mental well-being. International contributions to the study of positive mental health, pp. 299–329. : Springer.
4. Eriksson, M. (2017). The sense of coherence in the salutogenic model of health. In M. B. Mittelmark, S. Sagy, M. Eriksson, G. F. Bauer, J. M. Pelikan, B. Lindström, et al. (Eds.), *The handbook of salutogenesis* (pp. 91–96). Springer.

5. Seah, B., Tan, G. R., Eriksson, M., Wang, W., & Ramazanu, S. (2022). Re-orienting healthcare for healthy living communities: A qualitative exploration of nursing students utilising the salutogenic theory for community health practice. *Nurse Education Today, 119*, 105545. https://doi.org/10.1016/j.nedt.2022.105545
6. Mjøsund, N. H., Eriksson, M., Espnes, G. A., & Vinje, H. F. (2019). Reorienting Norwegian mental healthcare services: Listen to patients' learning appetite. *Health Promotion International, 2019*(34), 541–551. https://doi.org/10.1093/heapro/day012
7. Vinje, H. F. (2007). *Thriving despite adversity: Job engagement and self-care among community nurses*. Doctoral thesis, University of Bergen.
8. Langeland, E., Wahl, A. K., Kristoffersen, K., & Hanestad, B. R. (2007). Promoting coping: Salutogenesis among people with mental health problems. *Issues in Mental Health Nursing, 28*, 275–295. https://doi.org/10.1080/01612840601172627
9. Polhuis, K. C. M. M., Vaandrager, L., Soedamah-Muthu, S. S., & Koelen, M. A. (2021). Development of a salutogenic intervention for healthy eating among Dutch type 2 diabetes mellitus patients. *Health Promotion International, 36*(6), 1694–1704. https://doi.org/10.1093/heapro/daab020
10. Koelen, M. A., & Lindstrom, B. (2016). Health promotion philosophy and theory. In *Twenty-five years of capacity building: The ETC "healthy learning" process*. European Training Consortium in Public Health and Health Promotion (ETC-PHHP) and the Wageningen University, Health and Society Group.
11. Magistretti, C. M., Topalidou, A., & Meinecke, F. (2019). Sense FOR coherence—der Sinn FÛR Kohärenz: Annäherungen an ein mögliches Konzept. In C. M. Magistretti, B. Lindström, & M. Eriksson (Eds.), *Salutogenese kennen und verstehen. Konzept, Stellenwert, Forschung und praktische Anwendung* (pp. 119–135). Hogrefe.
12. Péres-Wilson, P., Marcos-Marcos, J., Morgan, A., Eriksson, M., Lindström, B., & AÍvarez-Dardet, C. (2020). "A synergy model of health": An integration of salutogenesis and the health assets model. *Health Promotion International, 36*, 1–11. https://doi.org/10.1093/heapro/daaa084
13. Lindström, B., & Eriksson, M. (2010). *The hitchhiker's guide to salutogenesis. Salutogenic pathways to health promotion*. Folkhälsan and The IUHPE Global Working Group on Salutogenesis.
14. Lundqvist, T. (1995). Chronic cannabis use and the sense of coherence. *Life Sciences, 56*, 2145–2150.
15. Sack, M., Kunsebech, H. W., & Lamprecht, F. (1997). Sense of coherence and psychosomatic treatment outcome. An empirical study of salutogenesis. *Psychotherapie, Psychosomatic, Medizinische Psychologie, 47*, 149–155.
16. Weissbecker, I., Salmon, P., Studts, J. L., Floyd, A. R., Dedert, E. A., & Sephton, S. E. (2002). Mindfulness-based stress reduction and sense of coherence among women with fibromyalgia. *Journal of Clinical Psychology in Medical Settings, 9*, 297–307.
17. Körlin, D., & Wrångsjö, B. (2002). Treatment effects of GIM therapy. *Nordic Journal of Music Therapy, 11*, 3–15.
18. Blomberg, J., Lazar, A., & Sandell, R. (2001). Long-term outcome of long-term psychoanalytically oriented therapies: First findings of the Stockholm outcome of psychotherapy and psychoanalysis study. *Psychotherapy Research, 11*, 361–382.
19. Yamazaki, Y., Togari, T., & Sakano, J. (2011). Toward development of intervention methods for strengthening the sense of coherence: Suggestions from Japan. In *Asian perspectives and evidence on health promotion and education* (pp. 118–132). Springer.
20. Langeland, E., Riise, T., Hanestad, B. R., Nortvedt, M. W., Kristoffersen, K., & Wahl, A. K. (2006). The effect of salutogenic treatment principles on coping with mental health problems. A randomized controlled trial. *Patient Education and Counseling, 62*(2), 212–219. https://doi.org/10.1016/j.pec.2005.07.004
21. Bringsvor, H. B., Langeland, E., Oftedal, B. F., Skaug, K., Assmus, J., & Bentsen, S. B. (2018). Effects of a COPD self-management support intervention: A randomized controlled trial. *International Journal of Chronic Obstructive Pulmonary Disease, 13*, 3677–3688. https://doi.org/10.2147/COPD.S181005

22. Højdahl, T., Magnus, J. H., Mdala, I., Hagen, R., & Langeland, E. (2015). Emotional distress and sense of coherence in women completing a motivational program in five countries. A prospective study. *International Journal of Prison Health, 11*(3), 169–182. https://doi.org/10.1108/IJPH-10-2014-0037
23. Guo, C., Deng, M., & Yu, M. (2024). Interventions based on salutogenesis for older adults: An integrative review. *Journal of Clinical Nursing, 33*(7), 2456–2475. https://doi.org/10.1111/jocn.17028
24. Fagermoen, M. S., Hamilton, G., & Lerdal, A. (2015). Morbid obese adults increased their sense of coherence 1 year after a patient education course: A longitudinal study. *Journal of Multidisciplinary Healthcare, 8*, 157–165. https://doi.org/10.2147/JMDH.S77763
25. Malm, D., Fridlund, B., Ekblad, H., Karlström, P., Hag, E., & Pakpour, A. H. (2018). Effects of brief mindfulness-based cognitive behavioural therapy on health-related quality of life and sense of coherence in atrial fibrillation patients. *European Journal of Cardiovascular Nursing, 17*(7), 589–597. https://doi.org/10.1177/1474515118762796
26. Merakou, K., Tsoukas, K., Stavrinos, G., Amanaki, E., Daleziou, A., Kourmousi, N., et al. (2019). The effect of progressive muscle relaxation on emotional competence: Depression–anxiety–stress, sense of coherence, health-related quality of life, and well-being of unemployed people in Greece: An intervention study. *Explore, 15*(1), 38–46.
27. Eriksson, M. (2019). Research supervision as a mutual learning process: Introducing salutogenesis into supervision using 'the collegial model'. *Health Promotion International, 34*, 1200–1206. https://doi.org/10.1093/heapro/day088
28. WHO. (1986). *Ottawa charter for health promotion: An international conference on health promotion, the move towards a new public health, November 17–21 1986*. World Health Organization.
29. Eriksson, M. (2007). Unravelling the mystery of salutogenesis. The evidence base of the salutogenic research as measured by Antonovsky's Sense of Coherence Scale. Doctoral thesis. Folkhälsan Research Centre, Research Report 2007:1. Turku, Folkhälsan.
30. Mittelmark, M. B., Sagy, S., Eriksson, M., Bauer, G. F., Pelikan, J. M., Lindström, B., et al. (2017). *The handbook of salutogenesis*. Springer.
31. Mittelmark, M. B., Bauer, G. F., Vaandrager, L., Pelikan, J. M., Sagy, S., Eriksson, M., et al. (2022). *The handbook of salutogenesis* (2nd ed.). Springer.
32. Johansson, L. M., Eriksson, M., Dahlin, S., Lingfors, H., & Golsäter, M. (2024). Healthcare professionals' experiences of targeted health dialogues in primary health care. *Scandinavian Journal of Caring Science, 38*, 231–239. https://doi.org/10.1111/scs.13216

Open Access This chapter is licensed under the terms of the Creative Commons Attribution-NonCommercial-NoDerivatives 4.0 International License (http://creativecommons.org/licenses/by-nc-nd/4.0/), which permits any noncommercial use, sharing, distribution and reproduction in any medium or format, as long as you give appropriate credit to the original author(s) and the source, provide a link to the Creative Commons license and indicate if you modified the licensed material. You do not have permission under this license to share adapted material derived from this chapter or parts of it.

The images or other third party material in this chapter are included in the chapter's Creative Commons license, unless indicated otherwise in a credit line to the material. If material is not included in the chapter's Creative Commons license and your intended use is not permitted by statutory regulation or exceeds the permitted use, you will need to obtain permission directly from the copyright holder.

Chapter 7
Salutogenesis in the Context of Learning Processes

Lenneke Vaandrager, Maria Koelen, and Laura Bouwman

Learning, Healthy Learning and Healthy Learning Environments

Learning is the process of acquiring new understanding, knowledge, behaviours, skills, values, attitudes and preferences through experience and practice. Learning is basic to human behaviour, and it is a continuing part of life. Learning already starts before birth and continues throughout life. The acquisition of knowledge and the development of understanding are essential parts of the learning process [1, p. 45–48]. Learning often is described as an active, intentional and cognitive process, but it can also result from an associative process taking place without active thinking. This is based on the so-called "law of effect" where an action that leads to a desirable outcome is likely to be repeated in similar circumstances (e.g. Pavlov, Skinner). The process of thinking about relations between cause and effect is essential in learning. Learning not only results from direct—personal—experience and one's own actions. Behaviour is often learned from observing actions of others and consequences of those actions. In his social learning theory, Bandura [2, 3] refers to this as vicarious learning or modelling. Learning through observation or other interactions may also be unintentional. For instance, when an unexpected challenge occurs and people need to retrieve resources, they learn to actively apply the resources to deal with the challenge.

Learning takes place in every sphere of life, e.g. learning to play a musical instrument, a language, history, driving a car or how to prepare a meal. In the realm of this chapter, we focus on learning for health, or, more specifically, healthy learning. Partially based on Lindström & Eriksson [4], we consider healthy learning as a lifelong process where people and systems increase the control over, and improve

---

L. Vaandrager (✉) · M. Koelen · L. Bouwman
Health and Society, Wageningen University & Research, Wageningen, The Netherlands
e-mail: lenneke.vaandrager@wur.nl

health, well-being, and quality of life. Health literacy is an important concept for healthy learning but not the only concept. Health literacy broadly refers to the ability to "gain access to, understand and use information in ways which promote and maintain good health." Health literacy means more than being able to read pamphlets and follow prescribed health-seeking behaviours. It includes the ability to exercise critical judgement of health information and resources, as well as the ability to interact and express personal and societal needs for promoting health [5]. Being health literate is not enough for healthy learning. Being involved and empowered in the activities and decisions involving health where health is seen as a process over the life span and a resource for well-being clearly differs from the health-expert teaching knowledge and skills to special groups and the general public. Essential to learning is a healthy learning environment, which is a safe environment, characterized by clear structures and meaningful, empowering conditions that support individuals' and communities' cognitive, physical, psychological and social well-being.

Learning in the Context of Salutogenesis

Learning has been in focus in the field of salutogenesis from the very beginning up to date. A scholar who studied learning in the early days was Malka Margalit. She found that the SOC can facilitate learning processes among children and young people with learning disabilities [6]. The focus in her approach was to enrich language and basic learning skills and this worked out positively for both the SOC and for learning.

The first edition of *The Hitchhiker's Guide* (2010) distinguished healthy learning within health practice and within scholarly systems. In this chapter, however, characteristics and examples of salutogenesis in the context of learning processes in diverse environments are presented (see Fig. 7.1).

Salutogenic theory provides diverse guiding principles for shaping learning processes (based on our own work and on various authors of the second edition of *The Handbook of Salutogenesis* [4, 7–9]:

Learning processes are dynamic

- learning is part of a *life-long* development process that can be supported throughout all life stages;
- learning results from engagement in *multidimensional* learning involving mental, physical, social and spiritual aspects; hence, a diversity of strategies is applied to support this so-called embodied learning or learning with head, hands and heart [10];
- learning processes contribute to the development of SOC and SOC contributes to learning processes;
- the interlinked nature of *SOC elements* is addressed throughout learning processes, with meaningfulness considered as the key driving element;

7 Salutogenesis in the Context of Learning Processes

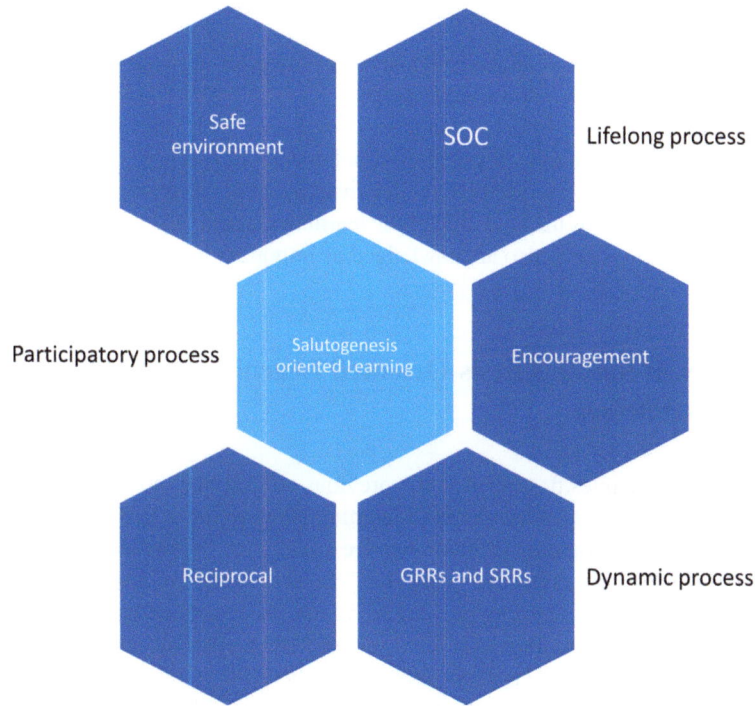

Fig. 7.1 Guiding principles for shaping salutogenesis-oriented learning

- *learning experiences* invite people to identify and apply resources in a way conducive to health and well-being, turning resources into GRRs.

Learning processes enable people to empower themselves

- learning is a dynamic and active process;
- *participatory* learning processes enable people to connect with their inner wisdom and gain control over their learning process;
- reflection on intentions, roles and outcomes supports development of *self-identity* which is important for the discovery of internal and external resources for learning.

Learning processes take shape in interaction and the learning process is reciprocal

- learning is done together through *observation, discussion* and other ways of interaction, with people mutually reinforcing each other in shaping what is learnt and how is learnt;
- learning processes support *emotional relatedness* among people or groups to incite effective learning.

Learning processes take place in a safe environment
- learning takes place in an environment that is safe for all, with clear *structures*, *meaningful conditions* and *social contacts;*
- learning processes respect *autonomy* so everyone feels safe to learn how to empower their cognitive, physical, psychological and social well-being;
- learning processes provide *positive feedback* and *encouragement*, with a focus on progress towards the ease-side of the learning process; when outcomes are not as expected, these are also acknowledged as important learning experiences.

Examples of Salutogenesis-Oriented Learning Within Different Settings

The guiding principles described in the previous section may not all apply to each learning process. Salutogenic-oriented learning takes shape in different forms and with different goals, ranging from knowledge transfer to eliciting motivation, building confidence and collective cohesion. In this section, five examples of salutogenesis-oriented learning are described that have applied (intentionally and unintentionally) the guiding principles in different ways. The examples include salutogenic-oriented learning processes within the settings of a health promotion university course and an international summer-course, healthcare, working in nature, a community and learning to eat healthy.

Salutogenesis-Oriented Learning in Scholarly Systems

The university MSc course 'Settings for Health Promotion' provides students with a well-defined structure and allows them to develop an agency over the learning process. This balance between agency and structure allows students to learn from successes and failures and develop SOC and turn resources into GRRs. A team of facilitator-teachers fosters mutual learning by expressing their doubts, uncertainties or questions rather than posing themselves as authorities in health promotion or salutogenesis. Students collaborate in groups of five with real-life case studies in settings such as a community, school or workplace. Classes, case-study work, individual- and group assignments enable students to learn and apply health promotion theory and methods in practice. The idea that their effort does not end up in the drawer, yet is used by case commissioners, contributes to students' perceived meaningfulness of their work. Students are challenged through reflective exercises such as individual expectation- and reflection papers with strict deadlines. There is also much free space to experiment with ideas, warm support from lecturers and moments of reflection and peer support. Learning takes place in various ways. First, students learn how to manage a diversity of perspectives, needs and wishes of research, practice and colleague students. They learn to make decisions in challenging situations, with the support of the facilitator-teachers that ensure that the situation remains relatively safe. Through this course, master students experience that learning involves much more than gaining knowledge on theory and prepares them for doing their MSc thesis and internship. High student evaluation scores show that they highly appreciate this salutogenesis-oriented learning process.

Based on similar salutogenic guiding principles, the European Training Consortium (ETC) Summer school is another example. In a mixed group of health promotion practitioners and researchers from different countries participants come together for 2 weeks to learn about theories, methodologies and application of health promotion. A team of tutors—also from various countries and backgrounds—create a safe and structured environment and provide input for discussion and exchange. The tutors also adapt to what they see and hear and try to be role models themselves. The participants get an opportunity to collaborate in an international context and practice intersectoral collaboration. Learning occurs through the process of (self)exploring, listening, reflecting and engaging in dialogue with other professionals (participants and tutors) with broad and diverse social and cultural backgrounds [11]. This also involves a lot of unexpected, unintentional learning that happens in a type of free space, warm interaction and support, group- and individual reflection.

Salutogenesis-Oriented Learning in Health Care

A second example is about how learning can be facilitated in a healthcare setting [12, 13]. Polhuis [12] developed and evaluated a program that aims to support people with type 2 diabetes mellitus in learning to eat and live healthily. The Salutogenic Framework was applied to develop a holistic, flexible, encouraging and supportive approach to individual and group learning. Characteristics of the program include a health-based approach in which people receive holistic, reflective and supportive guidance. Also, space for sharing and listening to each-others' personal stories enabled participants to feel seen and acknowledged and fosters emotional closeness. A mixture of learning strategies, ranging from providing disease and food-related information to self-examination, reflection, goal setting and stress management exercises were provided through individual and group sessions. The open, dynamic learning process with a focus on personal meaningfulness of healthful eating in everyday life enabled participants to actively engage in discovering resources that they can apply when facing challenges along their life course.

Salutogenesis-Oriented Learning in Nature

A third example concerns how a certain physical environment, in this case, nature, can offer many resources for learning. Hiemstra et al. [14] explored how people with limited capability for work (LCW) learn from working in a natural environment, in this case, the maintenance of nature reserves in The Netherlands. She found that natural environments offer a great diversity of (reoccurring) tasks and opportunities and are very rewarding because the results of efforts are immediately visible. Working in nature appeared to be relaxing because of its beauty and calmness and at the same time invited an active lifestyle as the work is physically demanding without being overdemanding. The work offered rhythm, structure, experiencing and learning in practice. The social and physical work setting provided opportunities to discover talents, develop ambitions and gain knowledge and skills in a safe setting—the whole context enabled employees to empower themselves. At the same time, it was not only nature—the rich environment of physical GRR and SRR—but also the forest managers who acted as wonderful facilitators of these learning experiences: they provided clear assignments, make employees feel seen and rewarded and provided continuous support within and outside the working setting. Or in other words, a social learning environment that was characterized by consistency, safety and support.

Salutogenesis-Oriented Intercultural Learning

The fourth example concerns a situation where an international student experienced difficulty in finding an internship in The Netherlands which was at first perceived as a problem by the University and the student herself until a community initiative welcomed her for a period of 4 months. The internship provider perceived her joining the initiative as an opportunity rather than a constraint (thinking in possibilities rather than in constraints) especially because they were struggling to involve groups of the community that hardly participated. The community initiative showed their struggles and were interested in learning that takes shape in interaction. The result was a salutogenic learning process on all levels: the student was able to build good and warm relationships with new community members using her own experience of being a foreign student in The Netherlands as a reference frame. She stepped into the situation with her own experiences and also was able to reflect on a meta level. She could bridge theoretical knowledge and practical knowledge and involved members of the initiative who learned a lot as well. A good example of reciprocal learning. Working in a Dutch language that was far from perfect appeared an important aspect that increased learning and facilitated the relationship with the new members.

Salutogenesis-Oriented Learning to Eat Healthy

The last example is about eating healthy. A salutogenic approach to nutrition promotion has the aim to facilitate a health-directed learning process through partaking in balanced, consistent and socially valued experiences. Experiences include learning about procedural knowledge of food such as gaining food literacy and healthful cooking practices. It is important that learning is socially embedded and covers all aspects of eating such as selecting, purchasing and preparing healthful food and meals. Such activities can be provided through school programmes as well as community cookery clubs. Another important socially valued learning setting is at home, where positive parent–child interactions at the dinner table and cooking with partners, family or friends support learning about eating well. Given that experiences with eating well have a cumulative learning effect throughout the life course, all crucial life stages besides childhood, such as leaving the nest, marriage and retirement should be considered [15].

Re-Defining Salutogenesis-Oriented Learning and the Role of the Facilitator

We started this chapter with a definition of learning, healthy learning and what learning means in the context of salutogenesis. Based on what we have described in this chapter we would like to define salutogenesis-oriented learning as

> A lifelong and dynamic process that provides meaningful experiences and invites people to actively participate in applying GRRs and SSRs, hence supporting SOC. Salutogenesis-oriented learning enables people to take control over their learning process through reflection, meaningful conditions, social interaction and encouragement within a safe environment.

The role of facilitators of salutogenesis-oriented learning processes is to take a pro-active role in providing a safe environment by holding space for participants to explore, identify and apply resources and reflect upon their learning journey. Holding space means the provision of a physical space (e.g. a clean room, a natural environment, a friendly community centre) and making participants feel comfortable, welcome and accepted through spoken and unspoken language. Also, there should be a balance between the structure (the steps, who is when to speak, how long) and the flow, requiring facilitators to watch time and ensure that the structure not become a limit. For example, by allowing more time for group sharing to maintain flow. Salutogenesis-oriented facilitators are also active learners who partake in mutual learning in which facilitators and learners are partners who take turns in the roles of facilitator and participant [16]. Mutual learning facilitates challenging cognitions, assumptions, ideas and skills of all learners in a respectful way, with the aim to develop new ideas. This type of learning is important in all settings, for instance between teachers and students, health professionals and clients, community workers and citizens. Eriksson [17] developed a tool that can be applied to facilitate mutual learning during PhD-supervision, the so-called 'collegial model of research supervision' (see Chap. 6). She stresses that mutual learning also should be fostered between members of the supervisory team.

Some authors like Koelen & Lindström [18, p. 34] have posed the idea that facilitators need a "Sense For Coherence." This implies that they can support the development of a sense of coherence among their participants. First, by recognizing which resources are needed and second, by providing these in a way that participants can identify and use these to initiate salutogenic mechanisms that contribute to a sense of coherence.

Conclusion

Salutogenesis-oriented learning is characterized by life-long, dynamic, empowering and reciprocal processes in safe and challenging contexts that support SOC. Social interaction, reflection and mutual learning are key for actively engaging in discovering resources for health, well-being and quality of life.

References

1. Koelen, M. A., & van den Ban, A. W. (2004). *Health education and health promotion*. Wageningen Academic Publishers.
2. Bandura, A. (1977). Self-efficacy: Toward a unifying theory of behavioral change. *Psychological Review, 84*(2), 191–215. https://doi.org/10.1037/0033-295X.84.2.191
3. Bandura, A., & National Institute of Mental Health. (1986). *Social foundations of thought and action: A social cognitive theory*. Prentice-Hall, Inc.
4. Lindström, B., & Eriksson, M. (2011). From health education to healthy learning: Implementing salutogenesis in educational science. *Scandinavian Journal of Public Health, 39*(6_suppl), 85–92. https://doi.org/10.1177/1403494810393560
5. WHO. (2021). Health promotion glossary of terms 2021.
6. Margalit, M. (1998). Loneliness and coherence among preschool children with learning disabilities. *Journal of Learning Disabilities, 31*(7), 173–180.
7. Mittelmark, M. B., Bauer, G. F., Vaandrager, L., Pelikan, J. M., Sagy, S., Eriksson, M., Lindström, B., & Meier Magistretti, C. (2022). *The handbook of salutogenesis* (2nd ed.). Springer.
8. Eriksson, E. (2019). Research supervision as a mutual learning process: Introducing salutogenesis into supervision using 'the collegial model'. *Health Promotion International, 34*(6), 1200–1206. https://doi.org/10.1093/heapro/day088
9. De Oliveira Olney, R., & Kiss, E. (2022). The application of salutogenesis to teaching and learning—A systematic review. *Developments in Health Sciences, 4*(3), 58–68. https://doi.org/10.1556/2066.2021.00035
10. Skulmowski, A., & Rey, G. D. (2018). Embodied learning: Introducing a taxonomy based on bodily engagement and task integration. *Cognitive Research, 3*, 6. https://doi.org/10.1186/s41235-018-0092-9
11. Vaandrager, L., Bonmati Tomas, A., Hofmeister, A., Alvarez-Dardet, C., Contu, P., Koelen, M., et al. (2022). Salutogenesis post-graduate education: Experience from the European perspec-

tive on the ETC-PHHP health promotion summer schools (1991–2020). In M. B. Mittelmark, G. F. Bauer, L. Vaandrager, et al. (Eds.), *The handbook of salutogenesis* (pp. 51–56). Springer. https://doi.org/10.1007/978-3-030-79515-3_7

12. Polhuis, K. (2023). *Flourish and nourish: Development and evaluation of a salutogenic healthy eating programme for people with type 2 diabetes mellitus*. Internal PhD, WU, Wageningen University. https://doi.org/10.18174/631882

13. Polhuis, K. C. M. M., Vaandrager, L., Soedamah-Muthu, S. S., & Koelen, M. A. (2021). Development of a salutogenic intervention for healthy eating among Dutch type 2 diabetes mellitus patients. *Health Promotion International, 36*(6), 1694–1704. https://doi.org/10.1093/heapro/daab020

14. Hiemstra, S. R., Naaldenberg, J., De Jonge, A., & Vaandrager, L. (2024). Salutogenic mechanisms in nature-based work: Fostering sense of coherence for employees with limited capability for work. *Health Promotion International, 39*, 1–12. https://doi.org/10.1093/heaper/daae127

15. Swan, E. (2016). *Understanding healthful eating from a salutogenic perspective*. PhD dissertation Wageningen University. https://edepot.wur.nl/378940

16. Koelen, M. A., & Lindström, B. (2005). Making healthy choices easy choices: The role of empowerment. *European Journal of Clinical Nutrition, 59*(Suppl 1), S10–S16. https://doi.org/10.1038/sj.ejcn.1602168

17. Eriksson, M. (2019). Research supervision as a mutual learning process: Introducing salutogenesis into supervision using 'the collegial model'. *Health Promotion International, 34*, 1200–1206. https://doi.org/10.1093/heapro/day088

18. Koelen, M. A., & Lindström, B. (2016). Health promotion philosophy and theory. In *Twenty-five years of capacity building: The ETC "healthy learning" process*. European Training Consortium in Public Health and Health Promotion (ETC-PHHP) and the Wageningen University, Health and Society Group.

Open Access This chapter is licensed under the terms of the Creative Commons Attribution-NonCommercial-NoDerivatives 4.0 International License (http://creativecommons.org/licenses/by-nc-nd/4.0/), which permits any noncommercial use, sharing, distribution and reproduction in any medium or format, as long as you give appropriate credit to the original author(s) and the source, provide a link to the Creative Commons license and indicate if you modified the licensed material. You do not have permission under this license to share adapted material derived from this chapter or parts of it.

The images or other third party material in this chapter are included in the chapter's Creative Commons license, unless indicated otherwise in a credit line to the material. If material is not included in the chapter's Creative Commons license and your intended use is not permitted by statutory regulation or exceeds the permitted use, you will need to obtain permission directly from the copyright holder.

Chapter 8
Salutogenesis in the Context of Work

Georg F. Bauer and Anja I. Lehmann

Relevance of Work for Health and Salutogenesis

Work plays a key role in human life. The working-age adult population spends a significant proportion of their life at work. Work not only supplies the financial means essential for living but also provides social resources in interaction with colleagues and customers. Furthermore, it offers structure to our daily lives, opportunities for continuous learning and a platform for making meaningful contributions to society. Thus, work is closely linked to health. This is illustrated by the consistent empirical findings that unemployment as well as poor physical and psychosocial working conditions impede our health. On the other hand, work is essential for well-being as exemplified in the WHO definition of mental health as "a state of mental well-being that enables people to cope with the stresses of life, realize their abilities, *learn well and work well, and contribute to their community*" [1]. Consequently, the workplace has been considered as a key setting for health promotion, which according to the Ottawa Charter should focus on the settings of everyday life, where people learn, work, play and love.

Because of this close link between work and health, from the beginning, the salutogenic model of health (SMH) has been systematically applied, tested and advanced in the context of work. These developments are summarized here and implications for promoting health at work will be presented in the last section.

G. F. Bauer (✉) · A. I. Lehmann
Center of Salutogenesis, Division of Public and Organizational Health, Epidemiology, Biostatistics and Prevention Institute, University of Zürich, Zürich, Switzerland
e-mail: georg.bauer@uzh.ch

The Original Model of Salutogenesis Applied to Work

Already early in the twentieth century in times of increasing division of labour (Taylorism), Kurt Lewin [2] noted that one's work and occupation is a two-faced matter: both a demanding, energy-draining *means for living* and a valuable source of a *fulfilling, purposeful (working) life*. Accordingly, also Antonovsky [3] stated regarding working life: "A distinction must be made between the elimination of stressors and the development of health-enhancing job characteristics." Despite his view that SoC is to a large extent static after an individual reaches adulthood, he still believed that also for older workers, sense of coherence "can be modified, detrimentally or beneficially, by the nature of the working environment."

Considering Antonovsky's writing on health-promoting factors at work [3], his original SMH can be specified and simplified for the context of work (Fig. 8.1). Job resources are part of the generalized resistance resources that allow for coherent work experiences, characterized by consistency, underload–overload balance and opportunities to participate in decision-making. Coherent work experiences help build up the general SoC of employees. SoC influences how employees perceive stressors at work and how they cope with them by mobilizing the appropriate job resources. Successful coping will move them towards the ease-end of the health continuum and further strengthen their general SoC. Finally, good health strengthens job resources and generalized resistance resources, just as stressors can weaken them. Such reciprocal mechanisms are depicted as dotted lines in Fig. 8.1.

The chapter on workplace in *The Handbook of Salutogenesis* [4, 5] summarizes the evidence that indeed as postulated by this model SoC: (a) is *influenced* by job demands and resources, (b) *influences* work-related health outcomes, such as burnout, stress symptoms and wellbeing, (c) acts as a factor linking job demands/resources and health outcomes (mediator) and (d) *influences how strongly* working conditions are linked to health outcomes (moderator).

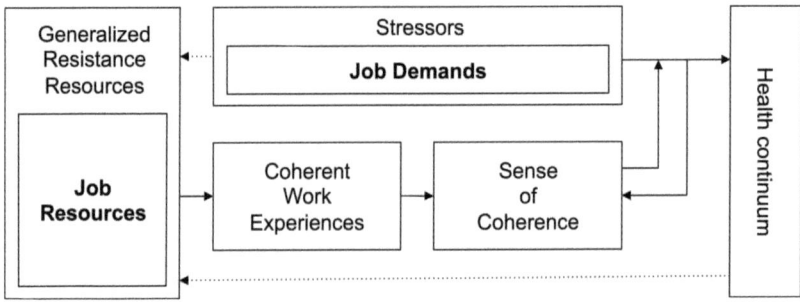

Fig. 8.1 Simplified specification of Antonovsky's original model of salutogenesis for the context of work [3]. Reprinted from [4], Fig. 31.1. https://doi.org/10.1007/978-3-030-79515-3_31, licensed under the terms of the Creative Commons Attribution 4.0 International License (http://creativecommons.org/licenses/by/4.0/)

The Job-Demands-Resources-Health Model as Work-Specific Model of Salutogenesis

The titles of Antonovsky's books [6, 7] show that the SMH in essence is a general stress model. It describes how SoC as a key personal resource helps to activate resistance resources to successfully cope with stressors—moving to the ease-end of the ease/dis-ease continuum. Since Antonovsky, stress research has been particularly advanced in the work context. Work-related stress research also emphasizes a balance between stressors or demands and resources as predicting health outcomes, as shown e.g. in the demand-control-support model or the effort-reward-imbalance models. During the last 20 years, these selective work-stress models have been broadened into the job demands-resources (JD-R) model, first published in 2007 [8]. It defines job demands broadly as negatively valued physical, social, or organizational aspects of the job that require sustained physical or psychological effort and are therefore associated with physiological and psychological costs. Through a path of health impairment, these demands lead to strain and finally exhaustion and burnout of employees. On the other hand, job resources are defined as positively valued physical, social, or organizational aspects of the job that are functional in achieving work goals, reducing job demands or stimulating personal growth and development. Through a motivational path, job resources lead to work engagement. In addition, the model postulates cross-over effects where job resources buffer the health-impairment process and job demands influence the motivational process. The relationships postulated by this model have been empirically proven in numerous studies. However, coming from the field of psychology, particularly on the motivational path, it focuses on psychological mechanisms and outcomes, missing the development of positive social and physical health.

In the field of health promotion research, in parallel, the health development model emerged [9] (Fig. 8.2). It shows the three interrelated dimensions of individuals' physical, mental and social health which develop through continuous interaction with relevant socio-ecological environments. This interaction can be observed from a pathogenic point of view, where risk factors lead to ill health and disease outcomes. Or from a salutogenic perspective, where resources buffer negative health outcomes of risk factors and directly promote positive health.

It was evident to combine the JD-R model with the health development model, given their parallelism. This led to the creation of the JD-R Health Model [10]. It postulates that JD-R are not only relevant for the outcomes of exhaustion and motivation but also for all three health dimensions. The health-impairment process is expanded to a "pathogenic path" leading to ill health defined as impaired physical, mental or social reproduction. Examples are musculoskeletal disorders, depressive mood and social exclusion. The motivational process is expanded into the "salutogenic path" of job resources leading to positive health defined as physical, mental and social fulfillment. Examples are energetic fitness, happiness and being embedded in harmonious relationships.

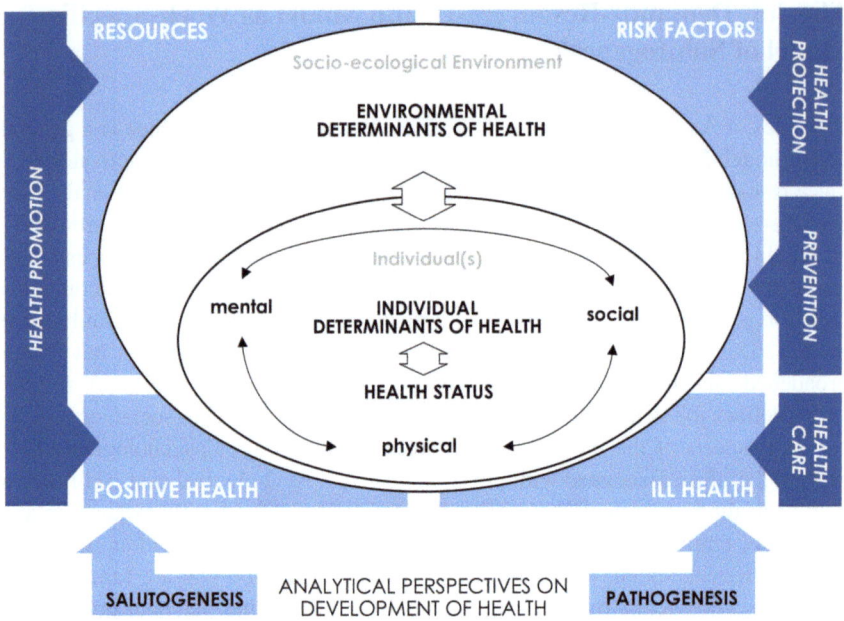

Fig. 8.2 Health Development Model [9]. (© The Author, 2006. Reproduced by permission of Oxford University Press. All Rights Reserved)

The original, empirical testing [10] showed that job demands such as time pressure or unclear roles are associated with a range of health outcomes including exhaustion, sleep problems and back and neck pain. Job resources such as holistic, complete tasks and good relationships with supervisors and colleagues were strongly associated with positive health outcomes including job satisfaction and commitment. Further, job resources indeed strongly buffered the negative impact of job demands. Finally, job resources were highly negatively related to the level of job demands—showing that they help to reduce avoidable demands in the first place. This triple power of resources together with evidence that job resources are more stable than job demands provides a strong argument for focusing more on job resources than job demands in worksite health promotion. It also suggests expanding the understanding of resources in the SMH beyond just providing generalized resistance against negative stimuli to also providing an immediate source of positive health experiences. So, growth and development can happen both by overcoming stressors (coping) and directly by experiencing a resource rich environment.

Figure 8.3 shows the original JD-R Health Model with its salutogenic and pathogenic paths leading to negative and positive health outcomes. However, to align it better with the SMH of Antonovsky, it also includes SoC and coherent work experiences as additional mediating and moderating paths of negative and positive health development.

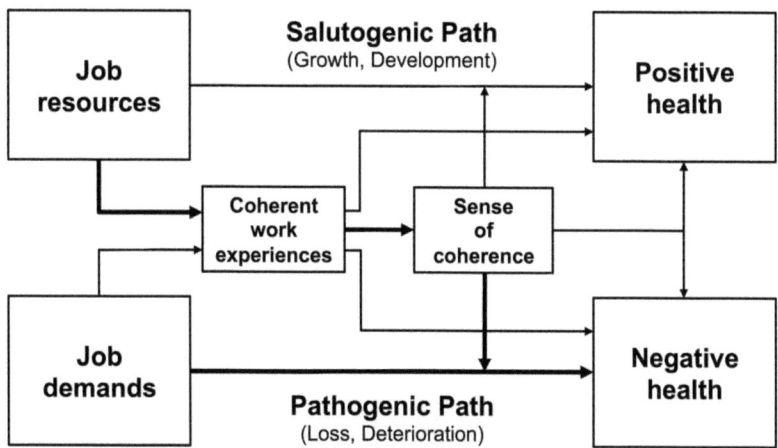

Fig. 8.3 JD-R Health Model, completed by coherent work experiences and SoC as mediating or moderating factors. Bauer GF & Jenni GJ, based on [10]. Adapted from Brauchli et al., 2015, Fig. 8.1. Some modifications were made. https://doi.org/10.1155/2015/959621, licensed under the terms of the Creative Commons Attribution 3.0 Unported License (https://creativecommons.org/licenses/by/3.0/) (**bold** = original salutogenic path) [4]. Reprinted from [4], Fig. 31.2. https://doi.org/10.1007/978-3-030-79515-3_31, licensed under the terms of the Creative Commons Attribution 4.0 International License (http://creativecommons.org/licenses/by/4.0/)

The Context-Specific Work-SoC as Indicator of Coherent Work Conditions

As mentioned above, the WHO Ottawa Charter states that health is created and lived by people within their everyday life settings (i.e. where they learn, work, play, love). This raises the question of how far people experience not only an overall sense of coherence as a "Global Orientation to Life" but also a context-specific sense of coherence or orientation towards the respective life settings. Accordingly, Bauer and Jenny suggested the concept of "Work-related Sense of Coherence" (Work-SoC), defined as the perceived comprehensibility, manageability and meaningfulness of an individual's current work situation [11, 12]. They consider Work-SoC as an interactional concept influenced by both the more stable, underlying general SoC of employees, as well as by their more fluctuating working conditions, particularly their job demands and job resources. Being in the same work situation, a person with a strong general SoC will perceive the work situation as more comprehensible, manageable and meaningful than a person with a weak SoC. On the other hand, it is to be expected that changes in the work situation will lead more immediately to changes in the Work-SoC than in the general SoC, as the latter is also influenced by other life settings and previous life experiences. This suggests Work-SoC as a more specific and sensitive measure of coherent work experiences than the general SoC.

To make it feasible to measure this concept routinely in the work-context, the authors developed and validated a nine-item Work-SoC scale. It asks employees to respond to the general question: "How do you personally find your current job and work situation in general?" by filling in bipolar pairs of adjectives, such as "manageable–unmanageable" or "structured-unstructured." As predicted, empirically, Work-SoC has been shown to be reciprocally related to both job-demands-resources and general SoC, as well as related to both negative and positive work-related health outcomes [4]. Thus, Work-SoC can be considered a feasible and valid indicator of salutogenic working conditions. The originally German Work-SoC scale has been translated into English, Norwegian, Finnish, French, Italian, Spanish, Dutch, Japanese, Chinese, and Czech, all available on the STARS Webpage (www.starssociety.org). Recently, an adapted version for employees with limited capability to work has been developed using a participatory approach [13].

Crafting—Employees Pro-actively Improving Their Working Life

In today's complex work environment, individuals face increasing demands on their skills and capabilities. Employees are expected to adapt flexibly to rapidly changing circumstances, technological advancements, global interconnectedness and the blurring of boundaries between work and personal life. They are also encouraged to actively shape and adapt their roles. From a salutogenic perspective, by recognizing and utilizing both personal and environmental resources, individuals can gain greater control over their health and make informed, proactive choices across work and personal domains.

Ideally, employees can fully express this creative and adaptive capacity, known as "crafting." Crafting is the pro-active behavior of shaping one's environment to meet personal needs, as opposed to merely reacting to external demands [14]. Unlike coping, which is reactive and focused on managing stress, crafting is a proactive approach that aims for long-term improvements in both work and personal life. Accordingly, it has been suggested that SOC might not only help in coping with stressors but can also support individuals in crafting because SOC helps "to (i) identify and understand one's psychological needs; (ii) engage in crafting efforts to achieve needs satisfaction and (iii) use available and build-up new resources during the crafting process" [15].

Applied to working life, a strong SoC is expected to support individuals in proactively engaging in job crafting by helping them recognize and understand their psychological needs, motivating them to shape their roles to meet these needs, and supporting them to draw on existing resources while developing new ones throughout the process. Evidence suggests that a strong SOC indeed enhances crafting both within and outside of work, which in turn positively influences well-being in both domains [15, 16]. This highlights the relevance of the crafting concept within the salutogenic framework. Nevertheless, it should be emphasized that external conditions (i.e. job resources) remain important to best enable this health-promoting mechanism between SoC and job crafting [16].

Salutogenic Work and Work Participation of Employees with Health Issues

A central principle of health promotion is its orientation toward health equity. Beyond considering the influence of the socioeconomic status on health, this also includes recognizing that employees with pre-existing health issues encounter additional barriers. Research shows that individuals with ill health often face greater challenges in fully participating in society and are particularly at risk of long-term work exclusion [17]. The salutogenic approach, with its focus on coping, adaptability and strengthening of resources of all people, wherever they are on the health continuum, can help mitigate these barriers. Evidence suggests that the provision of job resources — such as role clarity, control or social support — can significantly aid in retaining employees with a range of severity of multiple sclerosis within the workforce [18, 19]. Moreover, providing job resources can support employees with mental health issues and also in crafting their work in ways like employees without such conditions [20]. Another study with people working in nature who have a limited capability to work identified six salutogenic mechanisms contributing to their Work-SoC: "(i) having constructive working relationships, (ii) experiencing structure and clarity, (iii) receiving practical and emotional support, (iv) support in the creation of meaning, (v) experiencing and learning in practice and (vi) physical activity and (absence of) stimuli." [21].

Thus, promoting resourceful working conditions is beneficial for all employees, especially for those with existing health issues. In doing so, companies can actively contribute to a more inclusive and equitable work environment, ensuring that employees irrespective of their health status can remain and thrive at work.

Implications for Promoting Salutogenic Work

Considering the above knowledge, we can draw conclusions on how organizations can purposefully promote salutogenic working conditions. One strategy is to immediately refer to and build on the three dimensions of SoC. For example, in a clinic for psychiatry [22], the intervention researchers directly assessed which aspects of working life impeded or strengthened the comprehensibility, manageability and meaningfulness of nurses in this clinic. They used this assessment as a starting point for the participatory development of salutogenic interventions with these employees. However, for non-experts, the three dimensions of SoC are often rather abstract and do not resonate immediately with average employees. Thus, this approach of building on the SoC dimension is particularly promising for organizations that already apply salutogenesis to guide their client-related activities. For example, organizations offering their services in social care, health care or education within a salutogenic framework can easily refer to the three familiar dimensions of SoC. These dimensions can be used to reflect how their work currently affects both their own SoC and that of their clients. Such a dual client-employee approach allows us to identify strategies that ideally improve SoC of both groups simultaneously.

For strengthening the coherent work experiences and thus the general SoC of employees, Antonovsky suggested that organizations should specifically offer consistency, underload–overload balance, and opportunities to participate in decision-making. This is a practically useful framework for leaders of organizations to promote the health of their employees. Specifically, organizations and work teams can promote consistency through regular, clear communication of their goals and strategies. This is particularly important in the forefront of any planned organizational changes which require adaptations by the workforce. Regarding underload-overload balance, the JD-R Health model showed that in the work context, this is specifically about achieving a balance between job demands and job resources. The first step is regularly assessing and discussing this balance based on company-wide written surveys or moderated discussions within teams. The collective results make clear that addressing a potential imbalance is not the sole responsibility of the individual employee but also a core, shared task of a company and its members aiming for a healthy, sustainable workforce.

For developing interventions to improve the balance of JD-R, it needs to be considered that the most important JD-R arise and are experienced on the team level. Examples are role clarification, providing support and appreciation or addressing work overload. Here, team-leaders play a crucial role. A recent review showed that constructive leadership is strongly associated with an improved balance of JD-R, reduced burnout and increased engagement of employees, finally also leading to higher job performance [23]. Constructive leadership entails praising, supporting and caring for followers.

To boost such leadership, practical experience [24] shows that it is important that first team-leaders themselves can reflect and improve their own balance of JD-R. Through this self-application and experience, they tend to be highly motivated and competent to then moderate workshops to improve the JD-R balance in their own teams. The strategies for improving the JD-R typically cover changes on the individual, team, and organizational levels.

This systematic approach can be completed by also informally addressing the JD-R balance in regular team meetings or in the direct dialogue with individual employees. Individually, pro-actively crafting the own job also has a positive impact on work-related health. Thus, another recommended strategy is to encourage and support such crafting behaviors of individual employees and to provide the opportunity to exchange and support each other in crafting on the team level.

The participatory nature of all these interventions also addresses the third dimension of coherent work experiences, the participation in decision making. As demonstrated above, it can be expected that boosting coherent work experiences will first strengthen the work-related SoC, and by that in the long run also contribute to the overall SoC. Considering the diversity of needs and assets of a diverse workforce throughout such workplace interventions truly contributes to salutogenic, equitable working conditions.

The question remains how companies can be motivated to promote salutogenic work. It has been previously argued [25] that the shortage of qualified staff and the shift to an experience economy provides the unique opportunity to suggest to companies to provide the best possible experiences for both their customers and employees simultaneously. Such synergistic thinking and acting make it likely that companies sustainably address the quality of the fast-transforming working life, where salutogenesis can provide a clear compass.

References

1. WHO. (2022). *World mental health report: Transforming mental health for all*. World Health Organization.
2. Lewin, K. (1920). *Die Sozialisierung des Taylorsystems. Eine grundsätzliche Untersuchung zur Arbeits- und Berufspsychologie*. Verlag für Gesellschaft und Erziehung.
3. Antonovsky, A. (1987a). Health promoting factors at work: The sense of coherence. In R. Kalimo, M. A. El-Batawi, & C. L. Cooper (Eds.), *Psychosocial factors at work and their relation to health* (pp. 153–167). WHO.
4. Jenny, G., Bauer, G. F., Vinje, H., Brauchli, R., Vogt, K., & Torp, S. (2022). Applying salutogenesis in the workplace. In M. B. Mittelmark, G. F. Bauer, L. Vaandrager, J. M. Pelikan, S. Sagy, M. Eriksson, et al. (Eds.), *The handbook of Salutogenesis* (2nd ed., pp. 321–336). Springer.
5. Broetje, S., Bauer, G. F., & Jenny, G. J. (2019). The relationship between resourceful working conditions, work-related and general sense of coherence. *Health Promotion International, 35*(5), 1–12. https://doi.org/10.1093/heapro/daz112
6. Antonovsky, A. (1979). *Health stress and coping*. Jossey-Bass.
7. Antonovsky, A. (1987b). *Unraveling the mystery of health: How people manage stress and stay well*. Jossey-Bass.
8. Bakker, A. B., & Demerouti, E. (2007). The job demands-resources model: State of the art. *Journal of Managerial Psychology, 22*(3), 309–328. https://doi.org/10.1108/02683940710733115
9. Bauer, G. F., Davies, J. K., & Pelikan, J. (2006). The EUHPID health development model for the classification of public health indicators. *Health Promotion International, 21*, 153–159. https://doi.org/10.1093/heapro/dak002
10. Brauchli, R., Jenny, G. J., Füllemann, D., & Bauer, G. F. (2015). Towards a job demands-resources health model: Empirical testing with generalizable indicators of job demands, job resources, and comprehensive health outcomes. *BioMed Research International, 2015*(1), 959621. https://doi.org/10.1155/2015/959621
11. Vogt, K., Jenny, G. J., & Bauer, G. F. (2013). Comprehensibility, manageability and meaningfulness at work: Construct validity of a scale measuring work-related sense of coherence. *SA Journal of Industrial Psychology, 39*(1), 1–8. https://doi.org/10.4102/sajip.v39i1.1111
12. Bauer, G. F., Vogt, K., Inauen, A., & Jenny, G. J. (2015). Work-SoC – Entwicklung und Validierung einer Skala zur Erfassung des arbeitsbezogenen Kohärenzgefühls. *Zeitschrift Für Gesundheitspsychologie, 23*(1), 20–30. https://doi.org/10.1026/0943-8149/a000132
13. Hiemstra, S. R., Fleuren, B. P. I., de Jonge, A., Naaldenberg, J., & Vaandrager, L. (2024a). Sustainable employability of people with limited capability for work: The participatory development and validation of a questionnaire. *Journal of Occupational Rehabilitation, 35*(1), 1–11. https://doi.org/10.1007/s10926-024-10191-1
14. De Bloom, J., Vaziri, H., Tay, L., & Kujanpää, M. (2020). An identity-based integrative needs model of crafting: Crafting within and across life domains. *Journal of Applied Psychology, 105*(12), 1423. https://doi.org/10.1037/apl0000495
15. Tušl, M., De Bloom, J., & Bauer, G. F. (2022). Sense of coherence, off-job crafting, and mental Well-being: A path of positive health development. *Health Promotion International, 37*(6), daac159. https://doi.org/10.1093/heapro/daac159

16. Vogt, K., Hakanen, J. J., Jenny, G. J., & Bauer, G. F. (2016). Sense of coherence and the motivational process of the job-demands–resources model. *Journal of Occupational Health Psychology, 21*(2), 194–207. https://doi.org/10.1037/a0039899
17. Schuring, M., Burdorf, L., Kunst, A., & Mackenbach, J. (2007). The effects of ill health on entering and maintaining paid employment: Evidence in European countries. *Journal of Epidemiology & Community Health, 61*(7), 597–604. https://doi.org/10.1136/jech.2006.047456
18. Lehmann, A. I., Rodgers, S., Kamm, C. P., Mettler, M., Steinemann, N., Ajdacic-Gross, V., et al. (2020). Factors associated with employment and expected work retention among persons with multiple sclerosis: Findings of a cross-sectional citizen science study. *Journal of Neurology, 267*, 3069–3082. https://doi.org/10.1007/s00415-020-09855-w
19. Lehmann, A. I., Rodgers, S., Calabrese, P., Kamm, C. P., Wyl, V. V., & Bauer, G. F. (2021). Relationship between job demands-resources and turnover intention in chronic disease – The example of multiple sclerosis. *Stress and Health, 37*(5), 940–948. https://doi.org/10.1002/smi.3049
20. Lehmann, A. I., Kerksieck, P., & Bauer, G. F. (2025). Long-term development in job crafting in employees with and without mental health issues during COVID-19: The role of job resources. *Journal of Occupational and Organizational Psychology, 98*(1), e70002.
21. Hiemstra, S. R., Naaldenberg, J., De Jonge, A., & Vaandrager, L. (2024b). Salutogenic mechanisms in nature-based work: Fostering sense of coherence for employees with limited capability for work. *Health Promotion International, 39*, daae127. https://doi.org/10.1093/heapro/daae127
22. Idan, O., Braun-Lewensohn, O., & Sagy, S. (2013). Qualitative, sense of coherence-based assessment of working conditions in a psychiatric in-patient unit to guide salutogenic interventions. In G. F. Bauer & G. J. Jenny (Eds.), *Salutogenic organizations and change: The concepts behind organizational health intervention research* (pp. 55–74). Springer.
23. Pletzer, J. L., Breevaart, K., & Bakker, A. B. (2024). Constructive and destructive leadership in job demands-resources theory: A meta-analytic test of the motivational and health-impairment pathways. *Organizational Psychology Review, 14*(1), 131–165. https://doi.org/10.1177/20413866231197519
24. Bauer, G. F., & Jenny, G. J. (2018). Leadership and team development to improve organizational health. In K. Nielsen & A. Noblet (Eds.), *Organizational interventions for health and Well-being: A handbook for evidence-based practice* (pp. 237–261). Routledge. https://doi.org/10.4324/9781315410494
25. Jenny, G. J., & Bauer, G. F. (2023). New work – New interventions: Digital occupational health interventions and the co-creation of a human centered future of work. *Scandinavian Journal of Work and Organizational Psychology, 8*(1), 1–13. https://doi.org/10.16993/sjwop.185

Open Access This chapter is licensed under the terms of the Creative Commons Attribution-NonCommercial-NoDerivatives 4.0 International License (http://creativecommons.org/licenses/by-nc-nd/4.0/), which permits any noncommercial use, sharing, distribution and reproduction in any medium or format, as long as you give appropriate credit to the original author(s) and the source, provide a link to the Creative Commons license and indicate if you modified the licensed material. You do not have permission under this license to share adapted material derived from this chapter or parts of it.

The images or other third party material in this chapter are included in the chapter's Creative Commons license, unless indicated otherwise in a credit line to the material. If material is not included in the chapter's Creative Commons license and your intended use is not permitted by statutory regulation or exceeds the permitted use, you will need to obtain permission directly from the copyright holder.

Chapter 9
Salutogenesis in the Context of Society

Ruca Maass, Lenneke Vaandrager, and Jake Sallaway-Costello

Introduction

Although Aaron Antonovsky developed Salutogenesis as a sociological and systemic theory, the societal dimensions of Salutogenesis and coherence have rarely been discussed and explored [1, 2]. While the Sense of Coherence (SOC), is developed through an interactive learning process (see Chap. 7), the social structures which guide these processes have received little attention. This poses "The twin question…: How dangerous is our river?" and "How well can we swim?" [3, p. 14]. The last two decades have seen a shift in salutogenic thought towards how social structures create coherent and incoherent experiences [2, 4] and how we might protect and enhance those structures to build coherent societies.

In this chapter, we apply Salutogenesis to unravel "coherence" in the complex interplay among individuals, settings, and society-at-large, and try to outline how more salutogenic, coherent forms of development can be supported. To do this, we focus on the River (see Chap. 2) and how it affects the swimmer, their experience, performance and outcomes. We also expand the scope-of-interest into the positive aspects of life, and wonder: *How pleasant, adventurous, supporting and exciting can our river be? What opportunities does it offer to learn and enjoy swimming?*

R. Maass (✉)
Department of Neuromedicine and Movement Science, Norwegian University of Science and Technology (NTNU), Trondheim, Norway
e-mail: ruca.e.maass@ntnu.no

L. Vaandrager
Health and Society, Wageningen University & Research, Wageningen, The Netherlands

J. Sallaway-Costello
Division of Food, Nutrition & Dietetics, University of Nottingham, Nottingham, UK

© The Author(s) 2025
M. Eriksson et al. (eds.), *The Hitchhiker's Guide to Salutogenesis*,
SpringerBriefs in Public Health, https://doi.org/10.1007/978-3-031-89568-5_9

Fragmented Societies

Humans are born into a complex world, getting constantly bombarded with new stimuli that they need to make sense of. The society—the social, cultural and political environments—we grow up and live in has major influences on our understanding of the world and of our place in it [5, p. 118]. But what happens when this world we need to relate to is chaotic, ambiguous, and possibly eroding?

Looking at the world in 2024, perceptions of increasing fragmentation, polarization and conflict are eminent [6]. The fragmentation of social realities is linked to lower life satisfaction as well as decreased mental and self-rated health [7, 8], strained social relationships and declining trust in society and people in general [9]. Applying a salutogenic lens on such dividing perspectives often reveals examples of incoherence: conflicting aims, goals that are propagated but not pursued, and socially valued aspirations that are impossible to achieve from the individual's social position. For example, a majority of the global society is aware, and propagates, tackling climate change as a global goal—still, the establishment of climate-friendly energy production (such as wind parks) is regularly protested locally. These matters become even more complex, and give room for more incoherencies when taking more global developmental aims into account. For example, sanctioning car use in dense inner-city areas, but not in more affluent suburban areas, might fail to realize that this imposes an unjust economic strain on poorer people, while simultaneously claiming to advance social justice. Such incoherencies challenge comprehensibility and hinder coherent responses at the societal level. They might also contribute to perceptions that individual contributions are meaningless, spoiling the motivation to "conquer the challenge." Instead, efforts are made to re-establish perceptions of coherence: for better or for worse.

The Quest for Coherence

Making sense of this complex world is a core motivation and existential human need [5], [10, p. 78f]. Thus, individuals try to make sense of their experiences, no matter how chaotic they are. Meanwhile, "coherence" is a relational concept that describes how parts of a system relate to each other, including the individual navigating the system. To put it simply, individuals experiencing an incoherent world can try to re-establish coherence through two distinct approaches: they might either intensify their efforts to "explore the system" and "discover structure," or they might "limit the horizon" from which to draw meaningful experiences [11, p. 24]. This may result in dismissal of socially valued aspirations (e.g., opposing a career) or even withdrawal into homogenous social groups that offer alternative explanations about "what is going on" in the world, as seen in the rise of groups built upon adverse beliefs (such as denying climate change). The downfall of achieving coherence by shutting out perspectives is, of course, that the picture of the world we develop is partial: it does not reflect the complex reality to which we need to relate.

Developing a Sense of Coherence by shutting out conflicting information is reflected in what Antonovsky describes as a "fake" or "rigid" SOC: a SOC that might appear strong at first glance, but builds on few, inflexible understandings and coping strategies [11], p. 25f. Rigid understanding can in itself represent barriers to maintaining a coherent picture of the ever-changing world we live in [11, p. 144].

Incoherent perceptions of the world can result in incoherent agency, in which people aspire to goals by adverse means, or compromise their own interests in ways that worsen their situation in the long run: e.g. opposing better walkability in their neighbourhood due to perceived dependence on the car, without realizing that higher walkability will reduce this dependence, and enable cheaper and healthier ways to get around. Coherent agency describes the capacity to overcome challenges in ways that contribute to a salutogenic development (a movement towards health/health-promoting and health-supporting conditions) in the long run; for both the individual as well as for their surroundings. People are not mere subjects for environmental influences; they are active agents of their own life as well as co-creators of the social reality we live in: *"people are... proactive and have some choice in life; and... social institutions in all but the most chaotic historical situations can be modified to some degree."* [3, p. 15]. Societal coherence is created in circular interactions between individuals, groups, settings, organizations, communities, policies, institutions and actors of the whole society.

The Emerging Model of Societal Coherence

At the seventh International Conference on Salutogenesis "Everyday life and crises as opportunities for salutogenic transformation," Lodz, Poland in 2024, an emerging model for societal coherence was presented which explores these interactions from a salutogenic perspective (see Fig. 9.1).

Starting with systematic experiences linked to the development of a strong SOC and perceptions of coherence at the individual level, the model describes how these relate to policies and practices within and across various societal settings, including families, peer groups and local communities, and are shaped by wider societal contexts. To explore the issue, the establishment of wind parks in rural communities was chosen as an example: this often sparks conflicts in rural communities that are linked to incoherent experiences, such as a focus on global gains, while neglecting local environmental impacts. During the workshop, the interplay among people, settings and society was discussed through dialogue between the "corners" of the model, to explore solutions that combine matters of individual, settings and societal coherence.

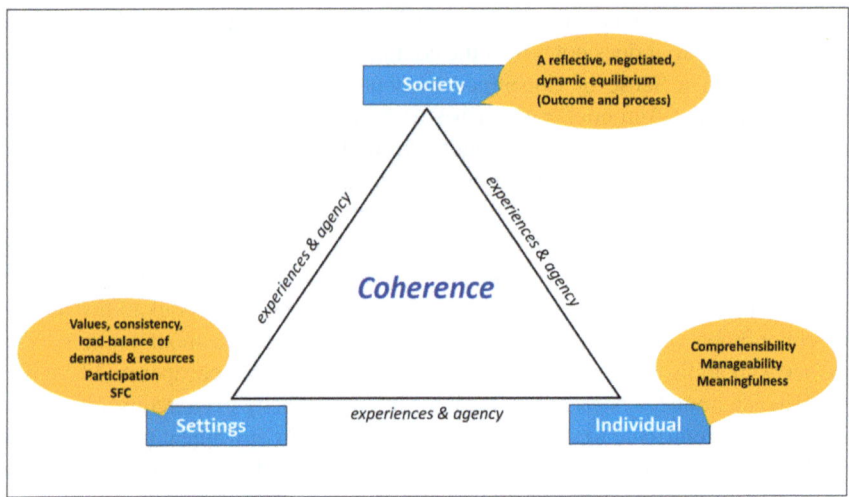

Fig. 9.1 The emerging model of societal coherence [12]. (Reprinted with permission of © Ruca Maass. All Rights Reserved)

How Society Shapes Our "Outlook on the World"

Lifting focus from matters of individual experiences to the societal conditions fostering a strong SOC increases the relevance of Salutogenesis for health promotion: *"The SOC, then, in turn would become a dependent variable, to be shaped and manipulated so that it in turn can push people toward health."* [3, p. 15]. The SOC is about making sense of the world and how we relate to natural and social realities we encounter; and implies *"a solid capacity to judge reality"* [5, p. 126]. Perceptions of coherence are based on experiences characterized by *"consistency, underload-overload balance, and participation in socially valued decision-making"* [3, p. 15]. There has been a tendency to focus on *"significant life events"* [5, p. 176], understood as major stressors or even "crisis," as drivers of the development of a strong SOC. Salutogenic interpretations of divorce [13], bereavement [14] and displacement [15], e.g. show how isolated social experiences can *"spark a series of unforeseen events"* [5, p. 176], interrupting daily life, forcing us to apply resources in new ways [16] and probably weakening SOC temporarily.

While Antonovsky pointed out the importance of such life events, he also emphasized the crucial role of *"repeated events,"* which help us to discover societal structure [11, p. 176]. Repeated experiences under slightly changing conditions enable us to apply resources in flexible ways: a core characteristic of a GRR [4, 17]. GRR's make major contributions to how we perceive challenging situations: how stressful, intriguing or exciting a challenge is experienced is heavily influenced by (a) our ability to cope with the stressor with available resources, and (b) the cultural and social interpretations of the situation [5, p. 72].

This illustrates the double-faced nature of stressors and resources and highlights the importance of how we understand a situation according to the meaning we find in it. For example, getting divorced might be associated with loss of status, failure and expecting hardship; or it might be experienced as taking control of the life situation and growing independence. This is dependent on individual as well as cultural and social conceptualizations of "divorce" [18]. Cultural narratives contribute to order and evaluate our experiences: they shape beliefs and identities, induce meaning, and guide how we make sense of the world [19]. Cultural narratives reflect and propagate societal values and norms, guiding the development of a *"repertoire of readily available automatic responses"* [11, p. 72], our "default" behaviour and coping strategies.

This implies that these are linked to cultural context, and even more specific cultural settings: how we cope with stressful situations at work is different from how we tackle them at home. For example, strong emotional displays are deemed inappropriate in a work conflict but may be integral for providing support to a family member in the home. Coping deemed "inappropriate" is, in turn, defined by cultural and settings-specific norms, as well as social position: a junior employee displaying anger triggers different reactions than an angry manager. Who we are in the societal systems has major consequences for how we perceive the world and our place in it.

Social Position and the Development of SOC

Antonovsky emphasized that life experiences, especially repeated experiences, depend on social position: *"The extent of such experiences is moulded by one's position in the social structure and by one's culture… with input from many other factors, ranging from gender and ethnicity to chance and genetics."* [3, p. 15]. Thus, the development and strength of SOC are linked to social identity, socioeconomic status, occupations, opportunities to participate in valued activities and feedback from others [11].

Social position impacts life events through two major mechanisms: different social positions imply different challenges and differences in the distribution of resources [5, p. 137]. In the River of Life (see Chap. 2), this may be seen as a range of structural factors that determine the speed of the water and the relative turbulence of the currents, representing systematic challenges with which a person must cope indefinitely, described as *"chronic life strain"* [20]. Such social positions align with the constitutional social determinants of health [21]. Socioeconomic advantage provides better opportunities and more secure access to resources, while socioeconomic disadvantage implies fewer opportunities (often linked to barriers to access societal relevant settings) and fewer resources to resolve challenges. Moreover, members of culturally defined minority groups may perceive the propagated life goals and norms of the dominant culture as irrelevant and demotivating, thus encouraging withdrawal from participation in societal settings.

Coherence in and Across Societal Settings

Every society is composed of a myriad of societal settings that provide opportunities to participate and enter a dialogue with our societal reality. Settings are defined as *"places or social contexts in which people engage in daily activities in which environmental, organizational, and personal factors interact to affect health and wellbeing* [22, p. 506]. The word "setting" can refer to different venues: schools, cities, islands, or hospitals, for example. Settings can have a formal organization (e.g. a workplace), a geographic situation (e.g. a community), similar conditions of life (e.g. senior citizens) or common values and preferences (e.g. religion). Nowadays, settings without well-defined physical boundaries such as online social networking sites are also considered settings in the context of health promotion [23].

Settings impose challenges, distribute resources, offer aspirations and contribute to experiences of structure, through inherent values, norms and narratives. Participation in (and across) societal settings provides opportunities to *"put feedback into the system and see how it reacts"* [5, p. 126], an important pre-requisite for understanding how society works (for the individual) and the development of adequate strategies to resolve challenges. Accordingly, access to relevant societal settings from which to draw experience emerges as a central resource in itself.

The understanding emerges that a **coherent setting** should provide opportunities to develop a coherent picture of the structure and mechanisms of the setting, and of how to manage relevant challenges in the setting with available resources. Participation in the setting can make a difference, for oneself, relevant others, or in the setting itself [4]. The degree to which values and aspirations are translated into policies, structures and resources that support these aims differs across settings and has a major impact on perceptions of coherence [5, p. 88]. Examples of incoherence in settings are easy to find, e.g. a school that aims to "give every child a good learning environment" but fails to provide children with special needs with the appropriate support.

However, the SOC – the **overall** outlook at the world – is developed based on experiences **across** settings. Transferring experiences, understandings and resources from one setting to another can be challenging and does not always yield desired outcomes: for example, nurturing social relationships in the family, in the workplace or in a digital community craves for different strategies. Being able to translate and apply resources across settings and situations allows for the flexible application of GRRs that characterizes, and reinforces, a strong SOC [11, p. 26].

To be described as a **salutogenic setting**, the setting should enable people to move towards health (not solely coherence) by processing life events and experiences in a reflexive way, linking life events to previous experiences, and looking at available capabilities and resources to find a solution. As health outcomes depend on the interplay between individual SOC and available resources, it is necessary to distinguish between resources and strategies supporting either coherence or health promotion aims: differences might occur [4, 24]. A "salutogenic setting" also implies that the setting itself is moving towards better health, and engages the people and the setting in a mutual upwards spiral; as e.g. proclaimed by the Healthy city network [25].

Can the Quest for Coherence Risk Salutogenic Developments?

Unravelling the relationships between coherent and salutogenic settings yields interesting insights into how individuals navigate complex realities and resolve incoherent experiences: by either further exploring the system to gain a deeper understanding, or by limiting their horizon, shutting out conflicting information and withdrawing into social in-groups sharing core beliefs about the world. While withdrawing into in-groups may aid coping in the short run, it may also support a non-salutogenic development at other levels: a strong sense of community and national coherence can contribute to individual health and wellbeing, but also to exclusiveness within and hostility towards other communities. Thus, while communities emerge as central coping resources and potentially coherent settings, too strong an emphasis on community coherence may contribute to societal fragmentation, especially in conflict situations [26, 27].

This is partly linked to a commitment to in-group narratives, often built on alternative interpretations of dominant cultural or political themes. For the individual, adopting an alternative explanation for the stressor can help to regain a sense of coherence in overwhelming situations. While commitment to in-group narratives can restore coherence in the short run and thus, support coping in the individual, it may also damage coherence and prevent a salutogenic development in the long run. Strong commitment to in-group narratives limits the range of societal settings, and experiences of transitions between them, from which to draw significant experiences, putting additional strains on in-group/out-group relations, enhancing in-group self-perceptions as being "morally superior" and legitimate in their aspirations [11, p. 21].

Such processes can be observed across the context of conspiracy theories and extremism, where in-groups are often defined by their beliefs in the (deviant) narratives of how the world "actually" works. Not coincidentally do conspiracy theories often utilize perceptions of challenges that are possibly overwhelming and hard to resolve from the individual position, such as people denying climate change and going far in accusing politicians, scientists and even climate activists of being part of a "climate conspiracy" that actively spreads lies and publishes false science to achieve a hidden (and unclear) agenda. This also illustrates the importance of a coherent picture of the world to develop coherent agency: Developing mistrust in politicians and environmental activists is only coherent if they are, as claimed, genuinely misleading the world on purpose. However, if the threat is real, as in climate change, then these strategies do not contribute to resolving the stressor; instead, they prevent effective societal action.

Antonovsky described such processes as developing a "fake" or "rigid" SOC, building on an incoherent picture of the world and resulting an inability to foresee outcomes: *"reality imposes itself and one is shattered"* [11, p. 25]. Looking at increasingly polarised debates across a range of topics, one might wonder if people that are strongly committed to in-group narratives adjust their beliefs in accordance with conflicting experiences. Raising levels of conflict within and between countries seem to indicate the opposite: a narrowing of valid perspectives by the devaluation of (ever more) perspectives that challenge the in-group narrative. To handle these challenges at a societal level implies further developing the **sense for coherence**; the ability to facilitate coherent experiences for others, beyond the individual and small-group level [28, 29].

Promoting Coherence for a Salutogenic Society

Exploring the interplay between individuals, groups and social settings, and developing strategies to support salutogenic developments at various levels is promising. In Antonovsky's words *"no society in history... managed to avoid structural limitations or... provide structural access to the goals it has propagated"* [11, p. 88]. Thus, all societies include aspects of incoherency, as well as opportunities to develop coherent perspectives. We have tried to unravel how developments that are both coherent and salutogenic can be pursued from different angles, addressing individuals, settings, or societal processes as such. However, during the development of the emerging model of societal coherence, possible downfalls and inherent challenges for a coherent and salutogenic society development emerged as well.

First of all, we do not know whether coherent settings contribute to or might damage the process of sense-making. Antonovsky proposed that a strong SOC would be constantly reinforced through seeking out challenges [5, p. 94f], while the absence of stressors and challenges is explicitly described as non-salutogenic [5, p. 86]. This implies that the individual process of meaning-making plays a crucial role in the salutogenic development. Could experiences that are too ordered, leaving no room for wonder and exploration, damage these processes? For example, entering a new work role and being given overly detailed descriptions of tasks might not facilitate the development of a sound understanding of one's role and responsibilities linked to the wider organizational contexts: the employee is limited to "ticking the boxes" without always understanding why. The same value of meaning-making can be seen in education (see Chap. 7), where didactic teaching approaches might make imparting knowledge more efficient, but in doing so remove any potential for meaning-making, thereby diminishing student engagement, participation and learning. By comparison, dialogical teaching approaches facilitate meaningful and effective learning by enabling student reflection on their relationship with the syllabus and providing further meaningful challenges by posing questions for future exploration. Is a certain amount of chaos and ambivalence necessary to ensure that the individual develops coherent perspectives, and becomes an active agent in their own life? How can we distinguish between "chaos" (the absence of structure) and incoherence (structures that do not work as intended)? And how to provide societal experiences that offer both the opportunity to navigate the system, but also the security of knowing it will lead somewhere?

Next, notions of a coherent society first and foremost point towards the internal logic of the societal system. "Coherence" does not imply any moral evaluation of strategies. Thus, a coherent society does not necessarily imply a nurturing, just, or morally good society. This is reflective of the explicit absence of cultural or social values in the original salutogenic theory [5, 11]. A society might be highly coherent in the way it pursues its goals, but simultaneously unhealthy by proclaiming adverse values such as self-centeredness, superiority, social injustice and short-sighted gains. Thus, in order to unravel coherence and Salutogenesis at a societal level, matters of societal values need to be considered.

Simultaneously, a strong emphasis on coherence combined with strong—maybe even exclusive—societal values makes for a "streamlined" society with little room for the development of individual perspectives or meaning-making. During the development of the emerging model of societal coherence, this danger of streamlining was even discussed in relation to one of history's least salutogenic societies: the German Third Reich (1939–1945). Could the Third Reich be described as a coherent society, given the strong societal values that were pursued in systematic manners throughout all of that society's institutions? It became obvious, however, that the opportunity to develop coherent perspectives in this society was limited, and strongly linked to social identity: for Jews, LGBTQ+ people, or political opponents, the most—and maybe only—coherent perspective was to leave the society.

Thus, a preliminary definition of a **coherent society** could be:

> a society which holds opportunities to develop coherent perspectives and agency for all its members, recognising a diversity of people as well as perspectives.

This points towards the importance of acknowledging, and supporting, diversity in perceptions of the world. Meanwhile, we cannot simply assume that a coherent society makes sound contributions to push people and communities towards health: coherent societies are not always salutogenic societies.

Salutogenic Societies

Applying the values of Salutogenesis and/or Health Promotion (such as equality, diversity, peace, good health and empowerment) to this definition of a coherent society might yield valuable insights into how a **salutogenic society** could be described and promoted. Salutogenesis is a theory of heterostasis, which in the context of society, values societal experience as inherently chaotic and subject to frequent, and often unpredictable, change. Antonovsky proclaimed that a salutogenic society would anticipate and facilitate for *"orderly change"* [5, p. 157], thus helping individuals and groups to resolve challenges coming with these changes. This notion is reflected in the emerging model of societal coherence (see Fig. 9.1), which envisions a salutogenic society as *"a negotiated equilibrium"* [12]. "Equilibrium" describes a state of balance between various (possibly opposing) forces in a way that supports perceptions of fairness. This definition emphasizes the relational nature of coherence and highlights the importance of developing a coherent picture by integrating, instead of shutting out, challenging perspectives. According to this definition, a salutogenic society could be envisioned as one where the various groups and individuals engage in constant negotiation and compromise to achieve a development based on a balance in their aspirations, interests and needs. Such a society would demand a certain willingness to compromise when developing solutions, as well as develop institutional support to facilitate negotiation and conflict resolution as part of their organization.

Facilitating **salutogenic dialogues** in which members of different groups exchange their thoughts and respective worldviews might be a way ahead towards a salutogenic society. A salutogenic dialogue is characterized by equity and a shared meaning-making process, in which various experiences are integrated into a holistic picture of the shared social reality—even beyond personal experience [4]. This makes it easier to understand why a resource works (or not) and can contribute to anticipating changes. For example, dialogues with people who are homeless or unemployed can give valuable insights into how societal resources work for groups experiencing hardship. From this starting point, the realization of shared perspectives and developmental goals is possible. Facilitating an ongoing dialogue across various levels and settings of society might be a valid strategy for achieving salutogenic development, with the broader goal of creating salutogenic societies. However, this demands that settings, groups and individuals engage in compromise and adaptation themselves, as well as evolving their societal structures, to conquer ever-new challenges in their path to salutogenic development. Some characteristics of salutogenic dialogues are shown in Table 9.1.

At this point, we want to draw attention back to the salutogenic side of life and emphasize the importance of framing positive goals and efforts as a common effort towards a better society, and not a last resort to stagger negative developments or avoid crisis [30, 31]. At a societal level, this requires engagement in the development of shared future visions that hold opportunities for various, and all, members of society.

Salutogenic-oriented health promotion efforts have by-large adopted the settings approach. Salutogenesis emphasizes social, personal and environmental resources for health, and provides conceptual orientation to settings approaches that address the health promoting capacity of everyday life contexts in which people "*live, work, play, and love*" [32, 33]. Settings approaches target groups and upstream resources for health and wellbeing, instead of individuals and single risk factors. For example, a health-promoting school offers an environment in which children can develop a feeling of comprehensibility, manageability and meaningfulness [34]. Salutogenic-inspired setting approaches often focus on the mobilization of internal resources and strategies linked to empowerment; and aim at facilitating a salutogenic development for the setting, as well as for the people in the setting. An important aspect would be to create opportunities for meaningful participation and impact in and across societal settings, to develop coherent agency and flexible strategies. Attempts

Table 9.1 Characterising salutogenic dialogue in the context of salutogenic societies

Non-salutogenic dialogue	Salutogenic dialogue
Meaning making privileges personal or in-group experiences	Meaning making, seen as a social process, is shared and includes diverse social groups and perspectives
Challenges ignored and narratives are constructed to explain away stressors	Challenges are exposed and explored, inform the common narrative and addressed as stressors
Diversity may be recognized, but only in the context of differing individual realities	Diverse social experiences feed into an integrated and shared social reality
Tension is resolved by protecting short-term perceptions of coherence by ignoring chaotic social realities and avoid potential adaptation	Tension is resolved by focusing on long-term developments; chaotic social realities are recognized to explore potential for orderly change
Tension between groups is resolved by ignoring or avoiding sensitive and possible conflicting issues.	Tensions between differing group interests are resolved by openly discussing sensitive and possibly conflicting issues in constructive ways (how can we resolve these in the future?)
Individualistic (or in-group) expectations override any interest in personal change for societal good.	Participants enter dialogue with a reflexive willingness to compromise and adapt their interests.
Avoiding discussion of challenges: focusing only on positive social experiences	Positive framings of challenges, embracing social issues as opportunities: talking about hard things in good ways

have been made to apply Salutogenesis as a guiding approach in (healthy) public policymaking, as well as with respect to structure collaboration activities in municipalities [35, 36]. Here, the framework of coherent experiences is applied in the planning and design of collaboration activities, anticipating that coherent structures can facilitate coherent processes as well as outcomes, at the individual as well as the collective level [36]. An important condition in this regard is to apply a salutogenic orientation: framing positive goals, instead of attempting to simply address challenges, enables collaborators to develop shared visions of the future and work towards achieving something good, rather than avoiding the negative [30], [31], [37]. However, while salutogenic theory states that a strong sense of coherence facilitates salutogenic development for individuals, we still need to unravel if and how a coherent setting can be described as a salutogenic setting, or support salutogenic development over time.

Concluding Remarks

Taken together, Salutogenesis addresses coherence as an interplay among societal, community and individual processes. To support societal coherence, dialogues within and between societal settings, as well as opportunities to influence and adapt societal settings from various positions seem crucial.

A salutogenic society's development craves not only for coherent processes and policies; but it also demands to development of positive visions of the future, the exploration of shared goals and the establishment of fair processes. Just as increasing social fragmentation is fuelled by perceptions of being confronted with overwhelming stressors and incoherent solutions; might a more salutogenic development be initiated by a focus on opportunities rather than crisis. Working towards positive visions (rather than avoiding a threat) is motivating and can contribute to identifying shared common aspirations and co-benefits (instead of emphasizing conflicts of interests and existing inequities). Engaging in the development of a shared, positive vision of the future society might even contribute to more salutogenic dialogues by providing a secure emotional base to resolve challenges (rather than just managing tension) and engage in common efforts to ensure that the River in fact is (increasingly) pleasant; intriguing and supportive for all swimmers.

References

1. Antonovsky, A. (1993). The sense of coherence as a determinant of health. In *Health and wellbeing* (pp. 202–211). Macmillan Education.
2. Eriksson, M., & Lindström, B. (2007). Antonovsky's sense of coherence scale and its relation with quality of life: A systematic review. *Journal of Epidemiology & Community Health, 61*(11), 938–944.
3. Antonovsky, A. (1996). The salutogenic model as a theory to guide health promotion. *Health Promotion International, 11*(1), 11–18.
4. Maass, R., Lindström, B., & Lillefjell, M. (2017). Neighborhood-resources for the development of a strong SOC and the importance of understanding why and how resources work: A grounded theory approach. *BMC Public Health, 17*, 1–3.
5. Antonovsky, A. (1979). *Health, stress, and coping. New perspectives on mental and physical well-being*. Jossey-Bass.
6. United Nations. (2024). *A new era of conflict and violence*. Accessed November 6, 2024, from https://www.un.org/en/un75/new-era-conflict-and-violence
7. Ku, B. S., Compton, M. T., Walker, E. F., & Druss, B. G. (2021). Social fragmentation and schizophrenia: A systematic review. *The Journal of Clinical Psychiatry, 83*(1), 38–87.
8. Hagedoorn, P., Groenewegen, P. P., Roberts, H., & Helbich, M. (2020). Is suicide mortality associated with neighbourhood social fragmentation and deprivation? A Dutch register-based case-control study using individualised neighbourhoods. *Journal of Epidemiology and Community Health, 74*(2), 197–202.
9. Dreyer, P., & Bauer, J. (2019). Does voter polarisation induce party extremism? The moderating role of abstention. *West European Politics, 42*(4), 824–847.
10. Mohr, G., & Willaschek, M. (Eds.). (1998). *Immanuel Kant: Kritik der reinen Vernunft*. Akademie Verlag GmbH.
11. Antonovsky, A. (1987). *Unraveling the mystery of health. How people manage stress and stay well*. Jossey Bass.
12. Maass, R., Meier Magistretti, C., Bauer, G., Contu, P., & Lindström, B. (2024). The missing link: Creating coherence between individual, group and society. IUHPE_ GWG Salutogenesis: 7th International Conference on Salutogenesis, Lodz, Poland, 19–20. June 2024.
13. Lindström, B. (1992). Children and divorce in the light of salutogenesis – Promoting child health in the face of family breakdown. *Health Promotion International, 7*(4), 289–296.

14. Magistretti, C. M., & König-Bachmann, M. (2019). It isn't just passed: A salutogenic perspective on bereavement care after stillbirth. *Clinical Obstetrics and Gynecology, 5*, 1–4.
15. Borwick, S., Schweitzer, R. D., Brough, M., Vromans, L., & Shakespeare-Finch, J. (2013). Well-being of refugees from Burma: A salutogenic perspective. *International Migration, 51*(5), 91–105.
16. Höltge, J., Mc Gee, S. L., Maercker, A., & Thoma, M. V. (2018). A salutogenic perspective on adverse experiences. *European Journal of Health Psychology, 25*(2), 53–69.
17. Maass, R. E. (2018). *The neighborhood as a salutogenic setting: How can Salutogenesis contribute to the development of strategies for promoting health and strengthening SOC through a focus on neighborhood-resources? Doctoral thesis*. Norwegian University of Science and Technology.
18. Malik, J., White, N., McLeod, H. J., & Hakim, C. (2024). The evolving experiences and impacts of divorce for women living in Palestine: a mixed-method narrative analysis. *Journal of Gender Studies, 33*(8), 1–16.
19. Rappaport, J. (2000). Community narratives: Tales of terror and joy. *American Journal of Community Psychology, 28*, 1–24.
20. Antonovsky, A. (1990). A somewhat personal odyssey in studying the stress process. *Stress Medicine, 6*(2), 71–80.
21. Dahlgren, G., & Whitehead, M. (2007). Policies and strategies to promote social equity in health. Background document WHO. Institute for future studies.
22. Poland, B., Krupa, G., & McCall, D. (2009). Settings for health promotion: an analytic framework to guide intervention design and implementation. *Health Promotion Practice, 10*(4), 505–516.
23. Loss, J., Lindacher, V., & Curbach, J. (2014). Online social networking sites—a novel setting for health promotion? *Health & Place, 26*, 161–170.
24. Maass, R., Lindström, B., & Lillefjell, M. (2014). Exploring the relationship between perceptions of Neighbourhood resources, sense of coherence and health for different groups in a Norwegian neighbourhood. *Journal of Public Health Research, 3*(1), 208.
25. World Health Organization. (2022). *How to develop and sustain healthy cities in 20 steps*. World Health Organization. Regional Office for Europe.
26. Sagy, S., & Mana, A. (2022). Salutogenesis beyond health: Intergroup relations and conflict studies. In M. B. Mittelmark, G. F. Bauer, L. Vaandrager, J. M. Pelikan, S. Sagy, M. Eriksson, et al. (Eds.), *The Handbook of Salutogenesis* (2nd ed., pp. 225–231). Springer.
27. Mana, A., Srour, A., & Sagy, S. (2019). A sense of national coherence and openness to the "other's" collective narrative: The case of the Israeli–Palestinian conflict. *Peace and Conflict: Journal of Peace Psychology, 25*(3), 226.
28. Koelen, M. A., & Lindström, B. (2016). Health promotion philosophy and theory. In *Twenty-five years of capacity building: The ETC 'Healthy Learning' process*. European Training Consortium in Public Health and Health Promotion (ETC-PHHP) and the Wageningen University, Health and Society Group.
29. Magistretti, C. M., Downe, S., Lindström, B., Berg, M., & Schwarz, K. T. (2016). Setting the stage for health: Salutogenesis in midwifery professional knowledge in three European countries. *International Journal of Qualitative Studies on Health and Well-Being, 11*(1), 1–12.
30. Maass, R., Lillefjell, M., & Anthun, K. S. (2021). Can salutogenesis contribute to prepare cities for climate change? In *City preparedness for the climate crisis* (pp. 93–105). Edward Elgar Publishing.
31. Maass, R., Kiland, C., Espnes, G. A., & Lillefjell, M. (2022). The application of Salutogenesis in politics and public policy-making. In M. B. Mittelmark, G. F. Bauer, L. Vaandrager, J. M. Pelikan, S. Sagy, M. Eriksson, et al. (Eds.), *The handbook of Salutogenesis* (2nd ed., pp. 239–248). Springer.
32. World Health Organization. (1986). *Ottawa charter for health promotion, 1986*. World Health Organization. Regional Office for Europe.

33. Mittelmark, M. B., Eriksson, M., Sagy, S., Pelikan, J. M., Vaandrager, L., Magistretti, C. M., Lindström, B., & Bauer, G. F. (2022). Salutogenesis for thriving societies. In M. B. Mittelmark, G. F. Bauer, L. Vaandrager, J. M. Pelikan, S. Sagy, M. Eriksson, et al. (Eds.), *The handbook of Salutogenesis* (2nd ed., pp. 635–638). Springer.
34. Jensen, B. B., Dür, W., & Buijs, G. The application of salutogenesis in schools. In M. B. Mittelmark, S. Sagy, M. Eriksson, G. F. Bauer, J. M. Pelikan, B. Lindström, et al. (Eds.), (pp. 225–235).
35. Lindström, B., & Eriksson, M. (2009). The salutogenic approach to the making of HiAP/healthy public policy: illustrated by a case study. *Global Health Promotion, 16*, 17.
36. Maass, R., & Lillefjell, M. (2022). Can the theory of Salutogenesis offer a framework to enhance policy coherence during policy development and implementation in municipalities? *Societies, 12*(1), 24.
37. Elliot, A. J. (2013). *Handbook of approach and avoidance motivation*. Psychology Press.

Open Access This chapter is licensed under the terms of the Creative Commons Attribution-NonCommercial-NoDerivatives 4.0 International License (http://creativecommons.org/licenses/by-nc-nd/4.0/), which permits any noncommercial use, sharing, distribution and reproduction in any medium or format, as long as you give appropriate credit to the original author(s) and the source, provide a link to the Creative Commons license and indicate if you modified the licensed material. You do not have permission under this license to share adapted material derived from this chapter or parts of it.

The images or other third party material in this chapter are included in the chapter's Creative Commons license, unless indicated otherwise in a credit line to the material. If material is not included in the chapter's Creative Commons license and your intended use is not permitted by statutory regulation or exceeds the permitted use, you will need to obtain permission directly from the copyright holder.

Chapter 10
From the Ottawa Charter to Planetary Health

Jake Sallaway-Costello, Claudia Meier Magistretti, and Bengt Lindström

A History of Health Promotion

This chapter uses salutogenic concepts to frame and reflect upon progress in health promotion, human rights, and global development. Salutogenesis was conceived as a meta-narrative of wellbeing: it was only at the end of his life that Antonovsky identified salutogenesis as the missing theoretical basis for health promotion [1], and advocated the use of this theory in health promotion for global development [2]. It is important that we appreciate that the movements and frameworks discussed in this chapter did not explicitly use salutogenic concepts to support health promotion. Instead, we use a salutogenic lens to appraise these developments and reflect upon how global development may be further progressed through new forms of salutogenic dialogue between people, groups, communities, and nations, in policy development processes and activism.

J. Sallaway-Costello (✉)
Division of Food, Nutrition & Dietetics, University of Nottingham, Nottingham, UK
e-mail: Jake.Sallaway-Costello@nottingham.ac.uk

C. M. Magistretti
Institute for Early Childhood Education, University of Graz, Graz, Austria
e-mail: claudia.meiermagistretti@uni-graz.at

B. Lindström
Norwegian University of Science and Technology (NTNU), Trondheim, Norway

Nordic School of Public Health, Gothenburg, Sweden
e-mail: bengtblind@hotmail.com

Origins of the Health Promotion Movement

Before health promotion became a human rights interest and a scientific discipline, it was an activist undertaking initiated by the second women's liberation movement in the 1970s [3]. This movement, based on women's rights, focused and redefined health as being crucial to political empowerment. Issues of sexual and reproductive self-determination were central to this. The "abortion issue" was seen as an aspect of social "body politics" [4], which manifested itself in the medical system like a burning glass. Sparked by the debate over the criminalisation of abortion, other reproductive health issues affecting pregnancy, birth, breastfeeding, mental health, menstruation and menopause were soon added. Activist criticism was directed at the defining power of medical experts, the view of women as weak, sick and suffering, and at the historical dispossession of women's healing knowledge by medicine. The central impetus of this discussion was self-help as a common practice by women for women, and the right to self-determination. New areas of knowledge were opened up, that would later become fundamental to the health movement, "new public health" [5] and health promotion. This included criticism of the medicalisation of bodies and lifestyles, patterns of women's use of the health system, discrimination against women as health and care workers, and the development of treatment methods appropriate to women [4].

By linking health and politics, the women's health movement did two important things: it challenged the medical definition of health and illness and undertook a critical analysis of women's illnesses such as frigidity, hysteria and depression. Second, it negated the conventional concept of politics, which was based on structures, institutions and objectified postulates, and declared the private sphere to be political, with living conditions and everyday life to be the arenas of health and self-determination. This conceptualisation, seeing health as being more than simply the absence of disease, lends to clear alignment with salutogenesis. The women's health movement was not explicitly based on salutogenic theory, but it showed a clear salutogenic orientation, through the focus on health and health outcomes, the rejection of a normative-static conceptualisation of health, the understanding of health as an interactive learning process, and the importance of resources and their development for the creation of health. This established the basis for a *"utopia of salutogenesis for women"* [6].

Health for All

Aligned with growing activist movements across the world, the establishment of the United Nations in the post-war period-initiated interest in the role of global goalsetting and cooperation in strengthening the relationship among health, human rights and global development. The global governance sphere at this time was firmly shaped by a longstanding prioritisation of economic development as the primary driver of all human achievement, largely influenced by US-aligned nations having prominent voice during the Cold War [7]. The evolution of the World Health Organization (WHO) challenged this assumption through the defining principles of the Health for All (HFA) movement [8, p. 409], which considered health to be a goal of economic development, not a tool for achieving it. This reversal of orientation, from good health for strong economies, to strong economies for good health, follows a salutogenic perspective that the medical focus on *"adding years to life"* should instead concern *"adding life to years"* [9].

A central conceptual achievement of the HFA movement was the early recognition of the distinct roles of clinical health care and public health activity, clarifying the WHO valuation of both the pathogenic and salutogenic perspectives on health. A statement made by the Director General of the WHO, Halfdan Mahler [8, p. 411] confirmed distinct and valuable roles for clinical medicine and public health, recognising that *"medical care alone cannot bring health to individuals."* The statement further called for public health policies and practices to bring *"a higher standard living,"* which could be interpreted to align with the salutogenic focus on Quality of Life as a pre-condition for health [10]. The HFA movement thereby established the role of global actors in setting goals for global health.

Ottawa Charter for Health Promotion

The role of the Ottawa Charter in formally establishing the field and practice of health promotion in 1986 has been introduced in Chap. 1 and cannot be understated. The Ottawa Charter rapidly became a game changer, shifting focus from the classic disease-risk approach of public health to that of a new health promotion action. It became the tool by which the contents of the HFA movement were implemented through the practice of health promotion. The Ottawa Charter itself was based on the core HFA valuation of preventative action and social change for health improvement but advocated its own set of principles and five action areas for achieving this [11], forming a comprehensive but easily understood integrated framework for health promotion. It created much enthusiasm in the field of public health, though some saw it as conflicting with traditional health literacy and behaviour change approaches. This was not the intention, as the Ottawa Charter was meant to create a concise form of common action connecting individual behaviour change with structural changes.

Regarding the WHO constitution (including the universal definition of health) there was a shift in thinking: in the Ottawa Charter, health was no longer seen as a "state," but as a lifelong process, driven by understanding how underlying social determinants such as socioeconomic status, gender (in)equality, warfare and transport shape health. It also advocated efforts to understand and modify these determinants, by both professionals and communities, as a driving force for change toward overall improvement of health. The WHO Euro Region was the first to establish a Health Promotion Division, initiating a series of workshops where the principles of the Ottawa Charter and the five action areas were scrutinised [12]. Already in 1987, it was discussed whether the WHO Health Declaration needed to be changed in order to complement the three recognised dimensions of physical, social and mental health, with a fourth dimension concerning spiritual and existential aspects of wellbeing [13]. Although not formally adopted in the universal definition of health, this dimension would have supported the salutogenic approach through its link with meaningfulness. However, despite advocacy by the Director General, the WHO was not ready to change its constitution [14] and this dimension remains unrecognised in the universal definition of health.

A major salutogenic achievement was that the Ottawa Charter generated a shift in the perceived outcome of the health promotion process: no longer was health a goal, but a process enabling the population to lead an active and productive life. This meant that the overarching focus was no longer only on reducing risk and preventing disease, but on creating wellbeing and enhancing quality of life in and for the population. This further aligns with the salutogenic interest in health-creating factors and processes, rather than the traditional disease-avoiding focus. The Ottawa Charter model, comprising the basic principles of health promotion and the five action areas, was to be seen as a coherent system focusing on an overall healthy and salutogenic development. Indeed, Antonovsky himself led a lecture at a WHO seminar in Copenhagen, Denmark (1992) in which the more explicit integration of salutogenesis into the Ottawa Charter was discussed, showing some intent for this focus [2].

One returning critique of this new approach to public health and health promotion was why it was not based on any theoretical foundation. This was later recognised as a strategic miscalculation, because theory would have given a structure to all forms of implementations [15]. What happened in reality was that instead of creating a coherent model for health promotion, where the Ottawa Charter principles would drive the five action areas, there were old existing groups, such as health education and health literacy scholars, that continued much as they did before, focusing only on individual behaviour change. We cannot underestimate how profound the legacy of this resistance to a shifting theorisation was in shaping current health promotion. Still today, individual behaviour change through health literacy is the dominant paradigm in health promotion policy and practice, despite decades of research consistently exposing its limited efficacy. Likewise, the action area "*building healthy public policy*" was largely interpreted in policymaking circles as referring to classic top-down health policy activity, and not as a participatory policy direction toward healthy development. This reluctance to shift towards salutogenic orientations of new concepts is attributable to already established traditional groups and organisations, finding it hard to reorientate their work towards the new health promotion action approaches [16]. In Europe, the Association of Schools of Public Health in the European Region (ASPHER) was delegated to set up working groups for each of the five action areas of the Ottawa Charter, to operationalise each of them into practical learning models. However, only one such group, focusing on The Lifestyle targets fulfilled the task. This group formed the European Training Consortium in Public Health and Health Promotion (ETC- PHHP) that still runs annual international training courses today [17]. Aaron Antonovsky was invited to teach on the second ETC course in 1992, at a time when salutogenesis was first being integrated into health promotion training. Thereafter, he was invited to WHO Euro to discuss how salutogenesis could serve as a theoretical base for health promotion [1]. Unfortunately, Antonovsky died shortly after this discussion and never saw how successful his model would become.

Millennium Development Goals

The approaching end of the twentieth century brought about renewed interests in global development, with clear influence from the health promotion perspective now defined by the Ottawa Charter. Global conferences held throughout the latter half of the century established a vast range of development targets [18]. Whilst many were pathogenic in orientation, seeking to eradicate disease, there was an emergence of salutogenic ambition, focused on achieving wellbeing through the strengthening of human rights and social equity. The development indicators of various global meetings were consolidated in the "*Shaping the 21st Century*" report [19], stressing a need for monitorable progress. At the same time, critique of the United Nations system throughout the 1990s raised concerns around top-down agenda-setting and incoherent processes [20], resulting in the "*We the Peoples: The Role of the United Nations in the 21st Century*" report [21]. Aligning the interests of development goals based upon human rights, and a call for UN reform to achieve human rights, the Millennium Development Goals (MDGs) established the role of the UN as a global goalsetting platform. Where the Ottawa Charter provided a guiding vision of health promotion action, the MDGs sought to assign measurable targets.

Presented as eight development goals, further broken into 21 measurable health or economic indicators, the MDGs were seen as achievable by the year 2015 [22]. From a salutogenic perspective, the MDG targets present a mixed offering of health-creating and disease-avoiding interests. For example, the focus of Goal 3 on the elimination of gender disparities in global education, seeks to strengthen GRRs by removing barriers to participation in learning. Conversely, the interests of Goal 6A in reducing transmission of HIV have been criticised for further pathologizing an already highly marginalised population [23], through excessive focus on testing and limited advocacy of anti-stigmatisation.

The combination of health and economic indicators as the measurable MDG targets raises the question of wealth as a proxy measure for well-being. One salutogenic perspective might argue that wealth enables meaningful life opportunities, and improves the manageability of life events, thereby strengthening the Sense of Coherence [24]. This conflicts, however, with the vision of salutogenic processes being rooted in and upholding human rights, misaligned, for example, with grievous human rights violations perpetrated by authoritarian nation states with high per capita income. Therefore, from a salutogenic perspective, in the absence of human rights, wealth cannot be assumed to promote, represent, or measure health. A major criticism of the MDGs is that the relatively random process by which they were developed, by cherry-picking ambitions from various global conferences, was led by wealthy nations [25]. This questions whether their creation can truly be valued as a form of salutogenic dialogue between nation states.

As the process of developing the MDGs was non-participatory and incoherent, they cannot be legitimately understood as salutogenic policy development. This explains the deeply uneven achievement of national-level implementation of the MDG targets. Those nations whos' interests were disregarded in the development of the goals, have typically made little progress towards achieving them, with many low-income nations having achieved none at all [26]. The absence of opportunities for meaningful participation, by people and nations, in the creation of the MDGs resulted in targets that were aspirational but unachievable, or in some national contexts, had no sociocultural relevance.

Sustainable Development Goals

Whilst they did establish a consensus framework for health promotion through global development, the MDGs provided only a limited basis for salutogenic progress. Criticism of the MDGs concerned their advocacy of downstream actions: targeting improvement of individual well-being, without modification of the structural and social determinants of health [27]. From a salutogenic perspective, this approach sees no shift in the River of Life from health disease to health-ease, but rather the placing of nets at the end of the River, attempting to catch people as they approach the waterfall and risk their wellbeing. This positioning frames the major areas of development, such as education, health care and social equity, as end goals rather than processes [27]. Salutogenically these are seen as generalised resistance resources (GRRs), which are understood not as ultimate goals, but pre-conditions for health. A framework was needed to evolve the end-point-focused MDGs into process-focused targets, while also shifting focus from individual action to collaboration. Following criticism of the top-down determination of the MDGs, the new framework would need to enable greater participation in global goalsetting, bringing about the creation of the Sustainable Development Goals (SDGs) in 2015.

The major differentiation between the MDGs and the SDGs was in how the two frameworks make sense of global inequalities and the role of nation states in resolving them. Where the MDGs established targets mainly aimed at less-developed economies, with the intent of (generally undelivered) assistance by more-developed economies, the SDGs had a more global and multidirectional feel [22]. This challenged the polarisation of nation states as development beneficiaries and benefactors, instead presenting a nexus of collaborators for global actions which recognised unequal access to resources, and disproportionate burden of risks. This may be seen as the early emergence of a somewhat salutogenic dialogue between nation states. The SDGs attempted a shift away from individual development for global health advocated by the MDGs, towards global development for individual well-being. As such, the salutogenic concept of the life situation is relevant to this transition. Antonovsky valued life situation as the sociocultural and historical positioning of the individual as a primary determinant of life stressors and access to GRRs [28]. By developing targets to which all nation states could contribute in unequal measure, the SDGs valued collective life situations, such as the degradation of poor nations through colonial extractivism, and the development of wealthy nations through environmental exploitation. It may be interpreted that the SDGs were the first such global development framework to consider a Collective Sense of Coherence for health promotion [29].

Planetary Health

By the arrival of the twenty-first century, the history of health promotion had been shaped by a participatory disconnect. In both the MDGs and SDGs, the social actors setting development goals were notably distanced from the people who were given responsibility for achieving them, and the settings where they happened. This is problematic. Antonovsky emphasised the role of participation in shaping meaningful life experiences. If excluded from the process of goal setting, the individual or community is unlikely to find meaning in the goals, and therefore experience little motivation to take action to achieve them. Antonovsky said, *"When others decide everything for us, when they set the task, formulate the rules, and manage the outcome, and we have no say in the matter, we are reduced to objects"* [29, p. 92]. The salutogenic approach thus stresses it is the *"taking part"* that enables and strengthens coherence [30]. Health-promoting development frameworks devoid of participation therefore cannot be truly salutogenic. Thus, following the courage of the women's health movement, the philosophy of the HFA movement, the vision of the Ottawa Charter, the goals of the MDGs and emergence of salutogenic dialogue between nations initiated by the SDGs, the most recent progress is the enabling of participation by the planetary health movement.

The central defining characteristic of planetary health is reciprocity and resonance: an understanding that people must be caretakers of the Earth, just as earth protects its inhabitants [31]. This shifts twentieth Century Western valuations of the environment from that of *planetary welfare*, in which the viability of ecosystems can be compromised when valuable for economic interests, to that of *planetary wellbeing*, which sees the needs of natural ecosystems as inherently non-negotiable. This post-anthropocentric understanding evolves salutogenic perspectives on ecosystems not only as physical GRRs but also as health subjects in their own right, with distinct personalities, recognising the finite availability of resources. In this sense, GRRs are not just to be accessed and used: they are to be protected and promoted.

It is important to note that the focus of planetary health on caretaking through reciprocity is not new. For millennia, Indigenous peoples have held meaningful, coherent relationships with natural ecosystems [32]. Many feminist activists who shaped the early origins of health promotion also identified parallels between anthroparchy and patriarchy [33], challenging the role of violence in diminishing human rights and preventing access to resources. The modern planetary health movement, regrettably (and inaccurately) often attributed to Western academic thought, must be understood as the delayed realisation of unheard and marginalised voices of groups with strong Sense of Community Coherence [34], muted by non-participatory attempts at global goalsetting throughout the twentieth century.

Planetary health forces a shift in perspective for health promotion. The threats of global ecosystem failure posed by climate change and other anthropogenic biohazards present health promotion with a challenge unlike those within the traditional scope of public health activity. There is no realistic or useful individualist approach to managing planetary threats, posing humanity with overwhelming and ever-more urgent questions for which traditional public health approaches have no real answers [35]. This can be understood through the salutogenic concept of load balance [30]. The urgent threat of imminent total global ecosystem collapse presents humanity generally, and health promoters specifically, with a challenge for which traditional tools and approaches are insufficient. This overload calls for humanity and the global health promotion community to identify new GRRs, and rethink how goals are set and achieved.

This process of realising new resources has commenced, through listening to those traditionally unheard voices that were excluded from processes of global goal-setting throughout the twentieth century. Through this, a process of *"planetary salutogenesis"* [36] has already begun, through the actions of grassroots activists, Indigenous peoples, youth changemakers, and other groups addressing the planetary determinants of health [37] through localised actions. This term refers not only to the application of salutogenic perspectives to planetary health, but further, describes the earth as a social actor in its own right, being subject to a process of salutogenic development. By transitioning health as a relational learning process [38] to planetary salutogenesis as a relational doing process [36], local actors become both the goal setters and the goal achievers, developing a Sense of Coherence that enables them to address planetary threats. In planetary salutogenesis, local groups translate overwhelming global challenges into local actions, giving context that empowers the development of new SRRs through strengthened comprehensibility of ecosystem threats. This provides a tangible base upon which to take action, enabling processes of joyful creation through which people can access existing GRRs and realise new ones, making achievement of planetary health more manageable. In doing so, overwhelming planetary threats have local significance, motivating participation in global change through local action, and engaging them in a process of meaning-making. Ultimately, planetary salutogenesis reconciles health as relational learning and doing, through new forms of salutogenic dialogue in which the deviant cases [29] of the planetary health movement can contribute to a growing "glocal" collaboration [36].

Table 10.1 A table summarising key shifts in participation in global development strategy

	Visionaries	Participation
Health promotion movement (1970s)	Feminist groups, civil rights activists, equality campaigners	Strong feminist advocacy of participation of people in decisions about their own health and bodies. Emerging inclusion in activist activity.
Health for All (1981)	WHO executive board, regional WHO advisors	Limited. Acknowledged global goal setting could not be done by medical doctors alone, but HFA visions set by WHO director.
Ottawa Charter (1986)	Health promotion professionals	Central focus of action area on community development. Valued local participation in policy making and health environment.
Millennium Development Goals (2000)	Various: UN chose goals from different global conferences	Participation shifts away from WHO high-level to national-level. Interests of wealthy nations prioritised; poorer nations trivialised.
Sustainable Development Goals (2015)	Member nations of the UN, with minor input from NGOs	Greater balance in participation of wealthy and poorer nations in design. Goals advocated action by diverse social and economic groups.
Planetary health movement (2020s)	Grassroots activists, indigenous peoples, youth changemakers	Focus on meaningful participation in acts of joyful creation. Valuing and listening to diverse ways of knowing for planetary health

Source: Authors

Conclusion

From the salutogenic perspective, the predominant shift in the creation and implementation of global development strategy for health promotion in the last half century has been an evolving reimagination of participation. While the integration of health actions and interests of local actors in the global context might initially seem challenging, we can see through the major global strategies of the late twentieth century, leading into the present day, that Antonovsky's vision of "taking part" [30] is, in fact, achievable. As noted in Table 10.1 the changing nature of participation indicates a form of salutogenic dialogue which is ever-strengthening, seeking to reconcile the interests of the goal-setters, with the actions of the goal-achievers.

References

1. Antonovsky, A. (1996). The salutogenic model as a theory to guide health promotion. *Health Promotion International, 11*(1), 11–18.
2. Lindström, B. (2022). Mileposts in the development of Salutogenesis. In M. B. Mittelmark, G. F. Bauer, L. Vaandrager, J. M. Pelikan, S. Sagy, M. Eriksson, et al. (Eds.), *The handbook of Salutogenesis* (pp. 5–9). Springer.
3. Daykin, N., & Naidoo, J. (2003). Feminist critiques of health promotion. In R. Bunton, R. Burrows, & S. Nettleton (Eds.), *The sociology of health promotion* (pp. 67–78). Routledge.

4. Kickbush, I. (1981). Health for all by the year 2000-people's health in people's hands. *The Jamaican Nurse, 21*(2), 32–33.
5. Ashton, J., & Seymour, H. (1988). *The new public health* (Vol. 1). Open University Press.
6. Meier, C. (1993). *Functioning and contradicting. Materials on the definition of women's health*. Institute of Social and Preventive Medicine, University of Bern, series of publications by the Working Group on Health Counselling.
7. Lee, J. (2022). Foreign aid, development, and US strategic interests in the cold war. *International Studies Quarterly, 66*(1), sqab090.
8. Mahler, H. (1981). The meaning of "health for all by the year 2000". *World Health Forum, 2*, 5–22.
9. Lindström, B., & Eriksson, M. (2011). From health education to healthy learning: Implementing salutogenesis in educational science. *Scandinavian Journal of Public Health, 39*(6_suppl), 85–92.
10. Lindström, B. (1994). *The essence of existence: On the quality of life of children in the Nordic countries-theory and practice in public health*. Nordic School of Public Health.
11. World Health Organization. (1986). *Ottawa charter for health promotion (No. WHO/HPR/HEP/95.1.)*. World Health Organization.
12. Catford, J. (2011). Ottawa 1986: Back to the future. *Health Promotion International, 26*(suppl. 2), ii163–ii167.
13. Mahler, H. T. (1987). World health for all: To be! *Bulletin of the Pan American Health Organization (PAHO), 21*(3), 306–313.
14. Peng-Keller, S., Winiger, F., & Rauch, R. (2022). *The spirit of global health: The World Health Organization and the 'piritual dimension' of health, 1946–2021* (p. 265). Oxford University Press.
15. Lindström, B., & Eriksson, M. (2006). Contextualizing salutogenesis and Antonovsky in public health development. *Health Promotion International, 21*(3), 238–244.
16. Eriksson, M., & Lindström, B. (2008). A salutogenic interpretation of the Ottawa charter. *Health Promotion International, 23*(2), 190–199.
17. Vaandrager, L., Bonmatí-Tomàs, A., Hofmeister, A., Alvarez-Dardet, C., Contu, P., Koelen, M., et al. (2022). Salutogenesis post-graduate education: Experience from the European perspective on the ETC-PHHP health promotion summer schools (1991–2020). In M. B. Mittelmark, G. F. Bauer, L. Vaandrager, J. M. Pelikan, S. Sagy, M. Eriksson, et al. (Eds.), *The handbook of Salutogenesis* (pp. 51–56). Springer.
18. Nayyar, D. (2013). The millennium development goals beyond 2015: Old frameworks and new constructs. *Journal of Human Development and Capabilities, 14*(3), 371–392.
19. OECD. (1996). *Shaping the 21st century: The contribution of development co-operation*. OECD Publishing.
20. Lee, T. (2010). The rise of international nongovernmental organizations: A top-down or bottom-up explanation? *Voluntas: International Journal of Voluntary and Nonprofit Organizations, 21*, 393–416.
21. Annan, K. A. (2000). *"We the peoples": The role of the United Nations in the 21st century*. No Title.
22. Sachs, J. D. (2012). From millennium development goals to sustainable development goals. *The Lancet, 379*(9832), 2206–2211.
23. Woodling, M., Williams, O. D., & Rushton, S. (2016). New life in old frames: HIV, development and the 'AIDS plus MDGs' approach. In *Framing global health governance* (pp. 62–76). Routledge.
24. Barnard, A. (2016). Sense of coherence: A distinct perspective on financial Well-being. *South African Journal of Economic and Management Sciences, 19*(4), 647–660.
25. Briant Carant, J. (2017). Unheard voices: A critical discourse analysis of the millennium development goals' evolution into the sustainable development goals. *Third World Quarterly, 38*(1), 16–41.
26. Larionova, M. (2020). The challenges of attaining the millennium development goals (MDGs). *International Organisations Research Journal, 15*(1), 155–176.

27. Loewe, M. (2012). Post 2015: How to reconcile the millennium development goals (MDGs) and the sustainable development goals *(SDGs)?* (No. 18/2012). Briefing paper.
28. Antonovsky, A. (1979). *Health, stress, and coping. New perspectives on mental and physical well-being*. Jossey-Bass.
29. Antonovsky, A. (1987). *Unraveling the mystery of health. How people manage stress and stay well*. Jossey-Bass.
30. Antonovsky, A. (1991). The structural sources of salutogenic strengths. In C. L. Cooper & R. Payne (Eds.), *Personality and stress: Individual differences in the stress process* (pp. 67–104). John Wiley & Sons.
31. Prescott, S. L., Logan, A. C., Albrecht, G., Campbell, D. E., Crane, J., Cunsolo, A. et al. (2018). The Canmore declaration: Statement of principles for planetary health.
32. Tu'itahi, S., Watson, H., Egan, R., Parkes, M. W., & Hancock, T. (2021). Waiora: The importance of indigenous worldviews and spirituality to inspire and inform planetary health promotion in the Anthropocene. *Global Health Promotion, 28*(4), 73–82.
33. Salleh, A. K. (1984). Deeper than deep ecology: The eco-feminist connection. *Environmental Ethics, 6*(4), 339–345.
34. Sagy, S. (2015). Coping, conflict and culture: The salutogenic approach in the study of resiliency. In *Resiliency: Enhancing coping with crisis and terrorism* (pp. 41–48). Ios Press.
35. Panico, F., Puga-Olguín, A., & Vargas-Madrazo, E. (2022). Organismic health in planetary emergency health situations: A transdisciplinary Salutogenic approach. *World Futures, 78*(8), 533–545.
36. Meier Magistretti, C., Sallaway-Costello, J., Fatima, S., & Hartnoll, R. (2021). People-planet-health: Promoting grassroots movements through participatory co-production. *Global Health Promotion, 28*(4), 83–87.
37. Redvers, N. (2021). The determinants of planetary health. *The Lancet Planetary Health, 5*(3), e111–e112.
38. Fox, N. J., & Alldred, P. (2016). *Sociology and the new materialism: Theory, research, action*. Sage.

Open Access This chapter is licensed under the terms of the Creative Commons Attribution-NonCommercial-NoDerivatives 4.0 International License (http://creativecommons.org/licenses/by-nc-nd/4.0/), which permits any noncommercial use, sharing, distribution and reproduction in any medium or format, as long as you give appropriate credit to the original author(s) and the source, provide a link to the Creative Commons license and indicate if you modified the licensed material. You do not have permission under this license to share adapted material derived from this chapter or parts of it.

The images or other third party material in this chapter are included in the chapter's Creative Commons license, unless indicated otherwise in a credit line to the material. If material is not included in the chapter's Creative Commons license and your intended use is not permitted by statutory regulation or exceeds the permitted use, you will need to obtain permission directly from the copyright holder.

Part III
An Outlook

Chapter 11
Critical Issues Related to the Salutogenic Theory and Its Implementation

Laura Bouwman and Lenneke Vaandrager

What We Measure

The Sense of Coherence (SOC) Scale

The Sense of Coherence (SOC) represents a perception of confidence in being well that is operationalised as '*a global orientation that expresses the extent to which one has a pervasive, enduring through dynamic feeling of confidence that one's internal and external environments are predictable and that there is a high probability that things will work out as well as can reasonably be expected*' [1, p. 123]. Several issues can be noted that relate to the dynamic and subjective nature of the SOC. Firstly, whether it is possible to assess a concept that represents continuous interactions between one's inner and outer world with a static scale. Perceptions are known not to be static but dependent on for instance mood, social and physical context. What exactly is it then that is measured through the SOC scale? Should SOC not be measured in multiple everyday contexts throughout life?

Another issue is of psychometric nature. Antonovsky stated that only one single total score, capturing all SOC scale questions should be calculated rather than separate scores for comprehensibility, manageability and meaningfulness. Psychometric evaluations of the SOC scale, however, have shown that the three components are not always as strongly related to each other as Antonovsky thought [2].

Also, the length of the scale as well as the items included have been questioned [3]. The appropriateness, comprehensibility and relevance of SOC scales appears to vary with life stage, socio-cultural background and literacy level. Naaldenberg et al. [4] for example found that the item of the SOC scale "Until now your life has had:

L. Bouwman (✉) · L. Vaandrager
Health & Society, Wageningen University & Research, Wageningen, The Netherlands
e-mail: Laura.bouwman@wur.nl

No clear goals or purposes at all ... Very clear goals or purposes" was perceived as referring to the future. Interviewees, elderly in this case, related these future goals to the context of occupation and work and therefore did not regard it as applicable to aging individuals who were already retired. The SOC scale also mainly appeals to cognitive capacity. This means that people with impairments that do not allow for reading or listening to questions might need SOC questions that are 'catchable' in pictures. However, there also have been recent and well-described attempts to adapt and validate the SOC scale to certain groups with disabilities such as people with limited capability for work [5].

Relation SOC and Health and Well-being Outcomes

Within salutogenic theory, health is comprised of four dimensions—the physical, mental, social and spiritual. There is substantial evidence for the predictive validity of SOC for especially in the mental dimension of health [6–8]. Also, evidence has become available on SOC and the spiritual health dimension [9] and diverse healthy life orientations such as healthful eating and physical activity [10–12]. For both healthful eating and physical activity, a dynamic interplay of health dimensions is found. Results of these studies show that if people feel more in control of life in general, they also start taking better care of themselves, in this case, eating healthier and exercising more.

Another issue is that within salutogenic theory, one's position on the health continuum is regarded as the health outcome, with the Sense of Coherence (SOC) as the principle predictive measure. This implies that SOC sketches people's movement along the health continuum, with an interplay of the four dimensions of health. Yet, the SOC scale is a set of questions, with this interplay of physical, mental, social and spiritual health being captured within the three elements of meaning, comprehension and management. Therefore, two questions arise. Firstly, whether a static scale is an optimal way of assessing this dynamic process. Secondly, what does the outcome, the SOC, represent? Is a 'strong SOC', which stands for 'I feel I have a meaningful life despite of/due to/with a disease' a basis for deciding whether action is needed? This could mean postponing (preventive) treatment until the situation worsens and the SOC weakens below the threshold. It raises issues regarding SOC being used as the single predictor of health status or whether other health outcomes should be added, as nicely expressed by Antonovsky himself:

> *"A salutogenic orientation, which does not in the least disregard the fact that a person has been diagnosed as having diabetes or is at high risk for breast cancer or shows signs of depression or has been given 2 weeks to live as a 'terminal cancer patient', of necessity, in asking, 'How can this person be helped to move toward greater health?' must relate to all aspects of the person."* [12]

Finally, there are other established theories with concepts that measure outcomes like SOC or its elements (see Chap. 3). The first edition of *The Hitchhiker's Guide* presented the umbrella with multiple theories and concepts contributing to the explanation of health. The similarities, differences and potential position within salutogenic theory have been described [13] for concepts such as flourishing [14] and self-transcendence [15]. Other relevant concepts such as empathy and learned hopefulness should receive further exploration [16]. Another unexplored concept at the individual level is intrinsic motivation, a concept within Self-Determination Theory [17]. Like the SOC element of meaningfulness, intrinsic motivation is about engaging in something out of a desire for agency over one's own life, out of love or contributing to higher goals. Further investigation of such concepts and their link, additional value and position within salutogenic theory will benefit the overall development of health promotion theory, research and practice.

Additional Measures Beyond SOC: Experiences, Mechanisms and Resources

The Ottawa Charter already stated that combining diverse but complementary methods or approaches is a principle of health promotion [18]. Looking at salutogenic theory, this particularly applies, since the origins of SOC lie within early childhood. During this life period, socio-cultural and historical living contexts bring stressors, resources, as well as life experiences that initiate a process of learning that shapes the SOC [19] (see Chap. 7). This learning process continues along the life course, with SOC levels indicating a certain level of feeling coherent at a moment in time. According to Antonovsky the SOC continues to develop up to the age of 30. Thereafter, SOC is relatively resistant to change yet temporary changes, and fluctuations around a mean may emerge [20, p. 124]. A recent review indicates that it is possible to strengthen SOC through interventions. However, the reviewed studies did not measure long-term outcomes [21]. In this respect, a longitudinal study [22] showed that SOC is quite stable and resumes its stability after stress (see Table 4.1 in Chap. 4).

To capture the dynamic, contextual nature of life, assessments should go beyond one moment in time and as well, capture the learning process that shapes SOC. By assessing SOC during multiple moments along the life course, insight is gained into how much SOC strengthens, weakens or whether it remains stable when people go through life stages. In research, explanations for this SOC dynamic may be found within major societal or personal challenges and opportunities that have functioned as learning experiences.

Secondly, additional measures are needed to more precisely capture how historical and structural contexts shape and are shaped by SOC. Insights from such measures can inform designs of salutogenesis-inspired societies (see Chap. 9). In the last decade, qualitative methods have been applied that often aim to unravel 'salutogenic mechanisms'. Examples include how working in nature drives functioning well at work [5], how experiences along the life course inspired healthful eating in later life [10] and how sports participation drives meaning [11]. In these studies, participants' life experiences in the past and present were collected with the aim of extracting the stressors they faced and resources they identified and applied to function well within society or in a certain setting.

How We Measure

Assessment Beyond the SOC Scale

Salutogenic theory has been applied in different ways, ranging from applying a salutogenic orientation, (parts of) the full model and a single focus on SOC [23]. Antonovsky argued that additional research methods such as life histories and in-depth interviews could provide better explanations for how SOC develops in different contexts [24]. Assessment tools beyond the SOC scale, specifically for qualitative investigations, have been developed, such as narrative tools. An example of such a tool is included at the end of this chapter.

Studies that aim to unravel 'salutogenic mechanisms' have been challenged by unclarity about the mechanism that links sense of coherence with movement on the health continuum. Mittelmark and Bauer [19] note that salutogenic theory poses that 'SOC helps to mobilise GRRs when faced with stressors by either (1) avoiding the stressor, (2) defining it as a non-stressor, (3) managing/overcoming, (4) leading to tension that is then, managed with success or (5) unsuccessfully managed tension; and that these outcomes influence one's movement on the ease-disease continuum', however, that it remains unclear what this mechanism exactly entails. This appears to complicate the analysis of narratives, especially distinguishing between resistant resources and life experiences. For example, whether employer-employee interaction should be considered a life experience itself, formed within this specific life situation. Or should it be considered a resource that is mobilised through life experience [25].

Tools additional to the SOC-scale should allow for extracting resistant resources that are relevant for individuals, groups and societies with diverse backgrounds, in a variety of situations that occur throughout life. In the last section of this chapter, a first set of characteristics for tools that align with principles, values and concepts of salutogenic theory is provided.

Power Dynamics

In this section, we discuss issues that relate to power dynamics that may be at play when applying salutogenic theory. Safeguarding equality in both research and practice is a cornerstone of health promotion and salutogenic theory. However, some scholars and practitioners have applied what can be regarded as a traditional, health education approach in which power dynamics occur [26]. The first issue is whether SOC can be influenced and if so, who has decisive power on what to do and how. Secondly, who is to be held accountable when inhabitants of resourceful societies are unable to find solutions to manage stress and maintain health, due to constraints outside their control.

Who Is in Control?

Several studies show that the SOC can be modified and strengthened among different groups through health-promoting interventions [21] The question is then, who is to influence what? In research, participatory designs such as participatory action research (PAR) foster equal distribution of power among all involved. However, it appears not to be easy to actively involve everyone in everything in all stages of a project or programme. For example, in a study evaluating community participation approaches to promote health, it was found that in some cases, community members felt overwhelmed by the responsibilities given to them, leading to feelings of stress [27]. The study by Mjøsund et al. [28] also shows that clarifying and discussing perspectives, responsibilities and roles of users of health services is an important feature of participatory research.

In addition, the 'helper-syndrome' is still present among researchers and practitioners. Although salutary health promotion should not affect personal autonomy by respecting that not all people have the same preferences, some professionals have problems with taking a less controlling role. Such a role can take the form of a partner and resource for the community at hand, rather than being the expert who assumes to know what's best. Besides the issue that handing over control can be problematic within professions that operate within a biomedical paradigm, there is the question of who is to have decisive power? For instance, for decisions about resources and experiences that should become available within a setting or society. Should it be based on the preferences of those that 'succeed' (strong SOC), those that 'fail' (weak SOC) or both? What to decide when the first group values self-actualisation whereas for the second group, togetherness is key?

Another issue relates to the mechanism that underlies the development or maintenance of a strong SOC. Namely, whether everyone has the same capacity and opportunities to identify and apply resources in a way conducive to health and well-being. Or does socio-culturally established power relations hinder some and benefit others?

Blaming the Victim

The SOC construct can lend itself to explanations and interventions that are neglectful of the fact that people in poverty often have very limited control over their circumstances [23]. For example, people living in a resource-rich society who have a weak SOC may be blamed for not making use of resources. It justifies expert control because obviously, when people are left to their own devices, they will naturally adopt an 'unhealthy' lifestyle [26]. This viewpoint is especially worrying as it has been expressed that the SOC not only depends on the individual and should not be used as a diagnostic tool (Personal communication between Dr. Avishai Antonovsky and Monica Eriksson 2024, see Chap. 4).

A 'blaming the victim' attitude can result from being part of the biomedical paradigm. Within this paradigm, human agency is idealised and implies that people can autonomously act upon their health, leaving little space for systemic and environmental influences. The influence of this paradigm is also visible in results of studies. For instance, research that applied salutogenic theory in the field of healthful eating has indicated more resources at the individual level than at the collective/societal level. For instance, studies on healthful eating indicate that self-efficacy, self-awareness, a reflective and positive attitude towards life, creativity and low doctor-oriented locus of control contribute to such eating [10, 29, 30]. These findings imply that research participants can more easily retrieve resources at the individual level. However, it may also reflect that people consider health as a merely individual responsibility, resonating with the biomedical perspective on health and its related factors.

Salutogenic theory, however, applies a systems approach to health, with the provision of experiences and resources that society does (not) provide as the foundation of SOC [31]. Studies that apply salutogenic theory, hence should acknowledge this interaction between individual capacities and societal contexts and view health as a collective, social responsibility. Actions based on the results of such studies then should address both structural societal changes and individual capacity building.

Further Steps

The issue of the multiple interpretations of 'salutogenic research and practice' indicates a need for clarity on how to use the theory to design health-promoting activities. The following guiding principles are indicated literature [21, 30]:

- Facilitate access and use of resistance resources
- Consider participants as a whole,

- Consider stressors and tensions as potentially health-promoting,
- Support individual and group learning processes,
- Ensure participants have active involvement and are allowed to influence the activity (active adaptation),
- Consider participants needs for emotional closeness with others,
- Consider participants needs for positive encouragement.

Further development of guiding principles in relation to other topics or settings may benefit the further development of salutogenic theory.

The development of methods that capture the dynamic, contextual nature of life that shapes SOC can be supported by extracting characteristics from salutogenic theory:

- *Investigate interaction*: with its roots in holistic, ecological health promotion, salutogenic theory assumes the world as dynamic and whole, with all beings connected and collectively constructing knowledge and understanding about their world [27]. This implies an assessment tool, including the type of analysis, that taps into this reciprocal interaction between people and context by investigating how participants make sense of a stressors, life-experiences and resources along their life-course within their historical and socio-cultural context.
- *Include multi-levels*: another consideration is that people live in multiple ecosystems, at individual, family-, group-, community- and population levels. Hence, an assessment tool should be sensitive to these different levels and how each of these may provide similar or different stressors, experiences and resources.
- *Consider multi-dimensionality at all levels:* the multi-dimensional nature of health is key to salutogenic theory and calls for an assessment tool that starts from a whole-person approach rather than single out particular aspects such as their physical status; also, SOC is a multidimensional concept and may include other elements besides meaningfulness, comprehensibility and manageability that should be investigated.
- *Active role of study participants*: assessment tools that are applied *with* participants rather than imposed on them means taking a participatory, collaborative approach in which participants' reflective capacity on what they find significant and meaningful is tapped into.
- Research that captures *both structural societal influences* and *individual capacity*.

> **Example of an Additional Method: Narrative Inquiry**
> As part of a larger study that aimed to understand healthful eating from a salutogenic perspective, a qualitative methodology known as narrative inquiry was used to explore life experiences and coping strategies that foster such eating. Narrative inquiry is defined as systematic listening to people's life stories. Stories were elicited through timelines, involving drawing and visually exploring life experiences to encourage participants to remember and reflect upon past experiences and make it easier to tell stories about their lives during the interviews. Participants also constructed a 'Food and Me' box which represented aspects that were important to them in terms of eating (e.g. objects, photo's, utensils). The box supported participants to reflect on their eating practices. Subsequently, 60 to 80 min interviews were held. First, participants were asked to discuss and explain their timelines chronologically from birth to the present. Then, describe key life experiences and turning points in relation to food and health. Third, the content of the Food and Me' box was discussed. The interviewer probed with questions when they wanted ideas or events to be described further. Interpretative phenomenological analysis (IPA) was applied to account for the world of participants and investigate events, processes and relationships. Informed by salutogenic theory, specific attention was paid to stressors, heuristics (strategies people employ in moments of uncertainty) and social and historical life paths. The study elicited insights in how healthful eating develops from exposure to individual- and context-bounded factors during childhood and adulthood and involves specific mental and social capacities including e.g. critical self-awareness, flexibility, craftiness and fortitude. Life-learning moments throughout the life course provided participants with opportunities to develop strategies that strengthened their agency and their capacity to overcome stressors. These findings inform holistic, life-long salutogenic-oriented nutrition promotion that, besides food and eating-specific factors, also enables general health-promoting practices such as mindfulness, critical thinking and stress management [29].[1]

References

1. Antonovsky, A. (1979). *Health, stress and coping*. Jossey-Bass.
2. Hochwälder, J. (2022). Theoretical issues in the further development of the sense of coherence construct. In M. B. Mittelmark, G. F. Bauer, L. Vaandrager, J. M. Pelikan, S. Sagy, M. Eriksson, B. Lindström, & C. Meier Magistretti (Eds.), *The handbook of Salutogenesis [internet]* (2nd ed., pp. 569–579). Springer.
3. Tušl, M., Šípová, I., Máčel, M., Cetkovská, K., & Bauer, G. F. (2024). The sense of coherence scale: Psychometric properties in a representative sample of the Czech adult population. *BMC Psychology, 12*, 293. https://doi.org/10.1186/s40359-024-01805-7

---

[1] Reprinted from [29] with permission from Elsevier.

4. Naaldenberg, J., Tobi, H., van den Esker, F., & Vaandrager, L. (2011). Psychometric properties of the OLQ-13 scale to measure sense of coherence in a community-dwelling older population. *Health and Quality of Life Outcomes, 9*, 37. https://doi.org/10.1186/1477-7525-9-37
5. Hiemstra, S. R., Fleuren, B. P. I., de Jonge, A., Naaldenberg, J., & Vaandrager, L. (2024). Sustainable employability of people with limited capability for work: The participatory development and validation of a questionnaire. *Journal of Occupational Rehabilitation, 35*(1), 1–11. https://doi.org/10.1007/s10926-024-10191-1
6. Eriksson, M., & Lindström, B. (2007). Antonovsky's sense of coherence scale and its relation with quality of life: A systematic review. *Journal of Epidemiology and Community Health, 61*(11), 938–944. https://doi.org/10.1136/jech.2006.056028
7. Schäfer, S. K., Sopp, M. R., Fuchs, A., Kotzur, M., Maahs, L., & Michael, T. (2023). The relationship between sense of coherence and mental health problems from childhood to young adulthood: A meta-analysis. *Journal of Affective Disorders, 325*, 804–816. https://doi.org/10.1016/j.jad.2022.12.106
8. Kieraitė, M., Novoselac, A., Bättig, J. J., Rühlmann, C., Bentz, D., Noboa, V., et al. (2024). Relationship between sense of coherence and depression, a network analysis. *Current Psychology, 43*, 23295–23303. https://doi.org/10.1371/journal.pone.0289203
9. da Silva Domingues, H., del Pino Casado, R., Palomino-Moral, P. Á., Martínez, C. L., Moreno-Cámara, S., & Frais-Osuna, A. (2022). Relationship between sense of coherence and health-related behaviours in adolescents and young adults: A systematic review. *BMC Public Health, 22*(1), 477. https://doi.org/10.1186/s12889-022-12816-7
10. Swan, E., Bouwman, L., Hiddink, G. J., Aarts, N., & Koelen, M. (2015). Profiling healthy eaters. Determining factors that predict healthy eating practices among Dutch adults. *Appetite, 89*, 122–130. https://doi.org/10.1016/j.appet.2015.02.006
11. Super, S., Hermens, N., Verkooijen, K., & Koelen, M. (2018). Examining the relationship between sports participation and youth developmental outcomes for socially vulnerable youth. *BMC Public Health, 18*(1), 1012. https://doi.org/10.1186/s12889-018-5955-y
12. Antonovsky, A. (1996). The salutogenic model as a theory to guide health promotion. *Health Promotion International, 11*(1), 11–18. https://doi.org/10.1093/heapro/11.1.11
13. Mjøsund, N. H., & Eriksson, M. (2021). Salutogenic-oriented mental health nursing: Strengthening mental health among adults with mental illness. In G. Haugan & M. Eriksson (Eds.), *Health promotion in health care—Vital theories and research* (pp. 185–208). Springer. https://doi.org/10.1007/978-3-030-63135-2_15
14. Mjøsund, N. H. (2021). A Salutogenic mental health model: Flourishing as a metaphor for good mental health. In G. Haugan & M. Eriksson (Eds.), *Health promotion in health care—Vital theories and research* (pp. 47–59). Springer. https://doi.org/10.1007/978-3-030-63135-2_5
15. Reed, P. G., & Haugan, G. (2021). Self-transcendence: A Salutogenic process for Well-being. Health promotion in health care—Vital theories and research. In G. Haugan & M. Eriksson (Eds.), *Health promotion in health care—Vital theories and research* (pp. 103–115). Springer. https://doi.org/10.1007/978-3-030-63135-2_15
16. Haugan, G., & Eriksson, M. (2021b). Future perspectives of health care: Closing remarks. In G. Haugan & M. Eriksson (Eds.), *Health promotion in health care—Vital theories and research [internet]* (pp. 375–380). Springer.
17. Deci, E. L., & Ryan, R. M. (2008). Self-determination theory: A macrotheory of human motivation, development, and health. *Canadian Psychology/Psychologie Canadienne, 49*(3), 182.
18. World Health Organization. (1986). *Ottawa charter for health promotion*. Ottawa charter for health promotion (who.int).
19. Mittelmark, M. B., & Bauer, G. F. (2017). The meanings of Salutogenesis. In M. B. Mittelmark, S. Sagy, M. Eriksson, G. F. Bauer, J. M. Pelikan, B. Lindström, et al. (Eds.), *The handbook of Salutogenesis [internet]* (pp. 7–13). Springer. https://doi.org/10.1007/978-3-319-04600-6_2
20. Antonovsky, A. (1987). *Unraveling the mystery of health. How people manage stress and stay well*. Jossey-Bass.
21. Langeland, E., Vaandrager, L., Nilsen, A. B. V., Schraner, M., & Meier Magistretti, C. (2022). Effectiveness of interventions to enhance the sense of coherence in the life course. In

M. B. Mittelmark, G. F. Bauer, L. Vaandrager, J. M. Pelikan, S. Sagy, M. Eriksson, et al. (Eds.), *The handbook of Salutogenesis [internet]* (2nd ed., pp. 201–219). Springer.
22. Piiroinen, I., Tuomainen, T.-P., Tolmunen, T., & Voutilainen, A. (2024). Meaningfulness and mortality: Exploring the sense of coherence in eastern Finnish men. *Scandinavian Journal of Public Health, 53*(1), 15–22. https://doi.org/10.1177/14034948231220091
23. Mittelmark, M. B., & Bauer, G. F. (2022). Salutogenesis as a theory, as an orientation and as the sense of coherence. In M. B. Mittelmark, G. F. Bauer, L. Vaandrager, J. M. Pelikan, S. Sagy, M. Eriksson, et al. (Eds.), *The handbook of Salutogenesis [internet]* (2nd ed., pp. 11–17). Springer.
24. Harrop, E., Addis, S., Elliott, E., & Williams, G. (2006). *Resilience, coping and Salutogenic approaches to maintaining and generating health: A review*. Cardiff Institute of Society, Health and Ethics.
25. Pijpker, R., Vaandrager, L., Bakker, E. J., & Koelen, M. (2018). Unravelling salutogenic mechanisms in the workplace: The role of learning. *Gaceta Sanitaria, 32*(3), 275–282. https://doi.org/10.1016/j.gaceta.2017.11.006
26. Gregg, J., & O'Hara, L. (2007). Values and principles in current health promotion practice. *Health Promotion Journal of Australia, 18*(1), 7–11.
27. Hogeling, L., Koelen, M., & Vaandrager, L. (2024). Community engagement in health promotion: Results from a realist multiple case study. *Health and Social Care in the Community* [Internet]. https://onlinelibrary.wiley.com/doi/epdf/10.1155/2024/2448483
28. Mjøsund, N. H., Vinje, H. F., Eriksson, M., Haaland-Øverby, M., Jensen, S. L., Kjus, S., et al. (2018). Salutogenic service user involvement in nursing research: A case study. *Journal of Advanced Nursing, 74*, 2145–2156. https://doi.org/10.1111/jan.1370
29. Swan, E., Bouwman, L., Aarts, N., Rosen, L., Hiddink, G. J., & Koelen, M. (2018). Food stories: Unraveling the mechanisms underlying healthful eating. *Appetite, 120*, 456–463. https://doi.org/10.1016/j.appet.2017.10.005
30. Polhuis, K. (2023). *Flourish and nourish: Development and evaluation of a salutogenic healthy eating programme for people with type 2 diabetes mellitus*. [internal PhD, WU, Wageningen University]. Wageningen University. https://doi.org/10.18174/631882
31. Antonovsky, A. (1991). The structural sources of salutogenic strengths. In C. L. Cooper & R. Payne (Eds.), *Personality and stress: Individual differences in the stress process* (pp. 68–102). John Wiley & Sons.

Open Access This chapter is licensed under the terms of the Creative Commons Attribution-NonCommercial-NoDerivatives 4.0 International License (http://creativecommons.org/licenses/by-nc-nd/4.0/), which permits any noncommercial use, sharing, distribution and reproduction in any medium or format, as long as you give appropriate credit to the original author(s) and the source, provide a link to the Creative Commons license and indicate if you modified the licensed material. You do not have permission under this license to share adapted material derived from this chapter or parts of it.

The images or other third party material in this chapter are included in the chapter's Creative Commons license, unless indicated otherwise in a credit line to the material. If material is not included in the chapter'-s Creative Commons license and your intended use is not permitted by statutory regulation or exceeds the permitted use, you will need to obtain permission directly from the copyright holder.

Chapter 12
Future Perspectives

Bengt Lindström, Monica Eriksson, Lenneke Vaandrager, and Georg F. Bauer

Monica Eriksson: Salutogenesis—A Whole World of Opportunities and Challenges

The production of this book has given new knowledge and broadened the scientific insights of the theory. Questions have been answered, while new questions have arisen. This personal view and vision of future research focuses on issues from two perspectives: *theory development* and *implementation* in practice.

In research clarification of concepts is an important step in the process of developing theories that are meaningful in the discipline and that make sense for people. It is not an endpoint, but a critical step in theory development. The concept of health has in this book been described as a process in a one-continua model (Antonovsky, see Chaps. 2 and 3) and in a two-continua model of mental health (Keyes, see Chap. 5). However, health is essential for life, but life is more than good health. This raises a vision to develop a four-continua model of well-being, consisting of spiritual, mental, physical and social elements. It is about QoL and optimal well-being, in the long run term, sustainability.

B. Lindström (✉)
Norwegian University of Science and Technology (NTNU), Trondheim, Norway

Nordic School of Public Health, Gothenburg, Sweden

M. Eriksson
Department of Psychology, Lund University, Lund, Sweden

L. Vaandrager
Health & Society, Wageningen University & Research, Wageningen, The Netherlands

G. F. Bauer
Center of Salutogenesis, Division of Public and Organizational Health, Epidemiology, Biostatistics and Prevention Institute, University of Zürich, Zürich, Switzerland

It has been obvious, that the dimensionality and the structural validity of the Sense of Coherence (SOC) still need further exploration (see Chap. 4). Today there is an increasing body of studies using Confirmatory Factor Analysis for exploring the dimensions of the SOC, showing that there are problems in understanding the items in the SOC scale. Further, recent research in Eastern cultures shows that the SOC scale seems to be more culturally sensitive than previously assumed, especially among older adults. This raises another question, is SOC more age-sensitive than we have assumed? Or is it about a sensitivity of a generation?

What does it really mean to be salutogenic, to think and to act salutogenically? Is it about "live the words" or "holding space," maybe. Research on this issue is limited. Some guiding principles are given, common for health promotion and salutogenesis, but need to be further explored. This leads to a need for more qualitative research to obtain a deeper understanding in different cultures and among various age groups. This can be achieved by using research methods where respondents and research subjects are involved as co-researchers and as active participating individuals.

Implementing salutogenesis in practice is more than only measuring SOC among individuals. It is more important to adopt the salutogenic guiding principles and think about how these can be systematically applied in a specific activity or context. This is especially relevant for workplace health promotion.

Finally, the lack of a *systematic* measurement of peoples' health resources is problematic. We have sufficient data on diseases and risks of falling ill, but we do not systematically measure resistance resources (see Chap. 3, Fig. 3.4, the umbrella). We can find research programs and health promotion projects; they are often temporary and of different lengths. They give us knowledge, but this is not enough to balance the risk approach with a resource approach to get the whole picture of peoples' health.

Georg F. Bauer: Advancing the Salutogenic Model of Health

In 2020, the Global Working Group (GWG) on Salutogenesis published a position paper on future directions for the concept of salutogenesis [1]. It identified four key conceptual issues to be advanced, including the overall salutogenic model of health. The paper recommends complementing the current ease/dis-ease continuum by an additional positive health continuum. The reason is that Antonovsky [2] defined the ease end of his ease/dis-ease continuum in a negative way, i.e. as the absence of pain, functional limitation, acute or chronic prognosis and health-related action implications. Thus, the ease/dis-ease continuum as a whole captures the domain of negative health, as it covers various degrees of absence of negatively valued, health-impairing aspects. However, already the WHO (1948) definition states that health includes both the absence of negative and the presence of positive aspects. Meanwhile, there is a broad literature capturing such positive aspects of health. This includes concepts like developing personal potential, well-being, well-functioning, self-fulfillment, pursuing a purpose in life, thriving or making a contributing to society. Now, one could just expand the definition of the ease-end of the single ease/dis-ease continuum accordingly—extending the continuum all the way into these positive aspects of health. However, this would create two problems. First, an ambiguity of definition: Being at ease could mean just being free of negative health (as postulated by Antonovsky). Or it could mean having moved anywhere into the positive health domain. Second, a single continuum assumes a straight, negative correlation between negative and positive health. On a single continuum, developing more dis-ease automatically removes one from experiencing positive health. And developing positive health automatically reduces dis-ease. However, developing a disease can go hand in hand with positive health development. Also, the dual continuum model of mental health and mental illness [3] suggests and empirically shows that these two continua are related in an orthogonal way.

The Job-Demands-Resources Model introduced in Chap. 8 distinguishes demands vs. resources as positively vs. negatively valued aspects of working life. In analogy, positive health could be defined as those aspects of health that are positively valued by individuals, which they would like to further approach and develop. Then, negative health captures those aspects of health that are negatively valued by individuals, which they would like to avoid or diminish.

Referring to the earlier health development model [4], the aforementioned position paper also recommends adding a path of positive health development leading directly from resources to positive health. This suggestion acknowledges that resources do not only play a key role as generalized resistance resource (GRR) and specific resistance resource (SRR) in helping to cope with or resist stressors and adversarial life situations but as growth resources, they can also support in approaching positively valued life goals and in personal growth and development. Such an expanded salutogenic model of health allows for universal studies promoting the full human health experience.

Lenneke Vaandrager: Future Research and Practice of Salutogenesis

In line with what is written in the second edition of *The Handbook of Salutogenesis* three future steps are required for future research and practice:

1. Sound application of the theory of salutogenesis in the health system, health promotion and other areas such as environmental development and sustainability, health governance and planetary health.
2. Theory development of the overall salutogenic model and continued emphasis on the study of quantitative and qualitative measurement tools.
3. Capacity building for the advancement of salutogenesis as an academic field [5].

The application of salutogenesis can especially gain from further operationalization in approaches in different settings and on different scale levels. In this book, we have included good examples in the area of societies, healthcare and workplaces. The second edition of *The Handbook of Salutogenesis* has included much more examples and the number of academic publications about so-called "salutogenic interventions or programmes" is rapidly growing in this field. In the area of environmental development (or protection) and sustainability there is still a world to win. Biodiversity might be an interesting starting point: when this increases(again) it creates meaningfulness to see different species flourish. One other popular development in this field is planetary health, which is widely embraced but has the danger of becoming dehumanized and only focused on risk management (outbreaks and disasters) and management of infectious diseases. Something happened during the COVID time, and increased inequalities, and has been unfortunate for the mental health of young people. Salutogenesis offers opportunities to unravel structural social factors such as resources for health and coherent dialogues. It can also serve as an orientation for governance when we want to improve the planetary health equity outcomes.

To advance salutogenesis as a theory rigorously there is a need to further develop and test the salutogenic model of health and salutogenic interventions that create, promote and restore well-being and the planet. Not as a recipe book but as an orientation that allows people to engage in this life-long, enriching and safe learning process. This also calls for the sound evaluation of our salutogenic programmes and policies: do these approaches have the intended impact, what works well and what works less well? What mechanisms are at play?

Salutogenesis also offers opportunities to work with research instruments that are salutogenic in themselves: that help to reflect on what is important for the quality of life and wellbeing and to try and learn from doing things differently in life. Or in other words, making participation in research is a pleasure and supporting the lifelong learning process.

For capacity building, we need salutogenic scholars all over the world to collaborate in exchanging how we teach salutogenesis in a salutogenic way. That requires a strong infrastructure, and international networks such as STARS (see Appendix) and the European Training Consortium for Public Health and Health Promotion ETC-PHHP to cherish communities that provide younger generations with training and education. To quote the final sentence of the second edition of *The Handbook of Salutogenesis*, "There is a tremendous diversity of opportunity of a salutogenic orientation to improve virtually all of society's well-being!"

Bengt Lindström

In 2026–2028 the UN and WHO will celebrate their 80th anniversary and the Ottawa Charter its 40th. They will be remembrances of the birth of the Human Right Movement in an idealistic time when hopes for a better future after the Second

World War were on top of the agenda. How is this connected to the Salutogenesis? The study that gave birth to Salutogenesis included women who were victims of the Holocaust but despite this still were able to carry on with a full and rich life of dignity. They were salutogenic.

The day the manuscript of the first edition of *The Hitchhiker's Guide to Salutogenesis* was finished, I started thinking of how it related to Douglas Adams's book *The Hitchhiker's Guide to the Galaxy* [6] which inspired me to give the title to this book. In Adams's book, a supercomputer is given the task of answering the ultimate question of the Meaning of Life. Mankind had though over time forgotten the original question and did not understand when the answer was simply the number 42. At the time it delighted me that the sum of the two original Orientation to Life Questionnaires (SOC 29 and SOC 13) was 42! That brought some laughs.

Now this time, I realise that Salutogenesis and Health Promotion 15 years ago mainly was concerned with only the Anthropocene perspective of the habitat on Earth. Things have changed dramatically! Today climate change and human-caused disasters, including grave violations of Human Rights have brought us to the ultimate question of the survival of the Planet and Life on Earth.

We now have an idea of how to bring it all back to one central mission. That is to create Coherence in and between people, habitats and utmost for life on Earth. The model and embryo for this is presented as the action and learning model in the chapter on Salutogenesis and Society. The question for the Future is simply how to create a coherent future for all and everything. This time let us not lose the Question!

However, we are not yet at the point of considering the Galaxies. My hope is that the Salutogenic perspective and response to the salutogenic question: "What Creates Health?" still is a challenge for the future hoping to create better conditions for Life on Earth for peace, harmony and coherence.

References

1. Bauer, G. F., Roy, M., Bakibinga, P., Contu, P., Downe, S., Eriksson, M., et al. (2018). Future directions for the concept of salutogenesis: A position paper. *Health Promotion International, 35*(2), 187–195. https://doi.org/10.1093/heapro/daz057
2. Antonovsky, A. (1987). *Unraveling the mystery of health: How people manage stress and stay well*. Jossey-Bass Publishers.
3. Keyes, C. L. M. (2014). Mental health as a complete state: How the Salutogenic perspective completes the picture. In G. F. Bauer & O. Hämmig (Eds.), *Bridging occupational, organizational and public health: A transdisciplinary approach* (pp. 179–192). Springer.
4. Bauer, G. F., Davies, J. K., & Pelikan, J. (2006). The EUHPID health development model for the classification of public health indicators. *Health Promotion International, 21*, 153–159.
5. Mittelmark, M. B., Bauer, G. F., Vaandrager, L., Pelikan, J. M., Sagy, S., Eriksson, M., et al. (Eds.). (2022). *The handbook of Salutogenesis* (2nd ed.). Springer.
6. Adams, D. (2012). *The hitchhiker's guide to galaxy*. Gollancz.

Open Access This chapter is licensed under the terms of the Creative Commons Attribution-NonCommercial-NoDerivatives 4.0 International License (http://creativecommons.org/licenses/by-nc-nd/4.0/), which permits any noncommercial use, sharing, distribution and reproduction in any medium or format, as long as you give appropriate credit to the original author(s) and the source, provide a link to the Creative Commons license and indicate if you modified the licensed material. You do not have permission under this license to share adapted material derived from this chapter or parts of it.

The images or other third party material in this chapter are included in the chapter's Creative Commons license, unless indicated otherwise in a credit line to the material. If material is not included in the chapter's Creative Commons license and your intended use is not permitted by statutory regulation or exceeds the permitted use, you will need to obtain permission directly from the copyright holder.

Appendix

Resources and Meeting Places of Salutogenesis: The Global Working Group, the Society, the Handbook and the Center of Salutogenesis

Georg F. Bauer
Center of Salutogenesis, Division of Public and Organizational Health, Epidemiology, Biostatistics and Prevention Institute, University of Zürich, Zürich, Switzerland, e-mail: georg.bauer@uzh.ch

Earlier Development of the Field of Salutogenesis

Immediately after the unexpected early death of Aaron Antonovsky in 1994, salutogenesis was primarily adopted and promoted in the Nordic countries by Bengt Lindström at the Nordic School of Public Health (NHV). He had been in close exchange with Antonovsky [1]. Health promotion and salutogenesis were included in the core programme of NHV, also introduced as a regular topic in the Nordic Health Promotion Research Conferences and the European Training Consortium (ETC). From 2008 to 2015, he organized international research seminars on Salutogenesis. In 2007, Bengt Lindström together with Maurice Mittelmark initiated the "Global Working Group on Salutogenesis" of the "International Union for Health Promotion and Education" (IUHPE) which he chaired until 2017. For broader dissemination, Bengt Lindström and Monica Eriksson published *The Hitchhiker's Guide to Salutogenesis: Salutogenic Pathways to Health Promotion* [2], available in English, Spanish, Catalan, French, Norwegian, Italian, German and Polish. They were also essential for running various centers for salutogenesis in Finland, Sweden and Norway.

Current Resources for Salutogenesis

Since 2017, Georg Bauer has been elected chair of the **Global Working Group on Salutogenesis**. The group includes over 20 selected experts in Salutogenesis around the globe covering diverse areas of applications of this concept. The group defined its mission as follows: "to advance and promote the science of salutogenesis (philosophy, theory, methodology, evidence) and thus to contribute to the scientific base of health promotion and the IUHPE" (www.iuhpe.org/index.php/en/global-working-groups). The group elects new members, seeking to include under-represented geographical areas, emerging research themes and those willing to work for our mission proactively. The group self-applies salutogenesis to its operation by considering principles of the Ottawa Charter; assuring inclusiveness regarding regions, gender, and age; following a coherent work plan, and voluntary engagement through joyful experience.

To further establish Salutogenesis as an interdisciplinary field, the Global Working Group published in 2017 [3] the first edition and in 2022 [4] the second edition of *The Handbook of Salutogenesis* with open access through Springer. By the end of 2024, the first edition achieved 2.88 Mil. accesses and over 1100 citations; the second edition, 1.25 Mil. accesses and 270 citations. *The Handbook* thoroughly shows the foundations as well as key advancements of the field of Salutogenesis. It also highlights that Salutogenesis has been applied to a broad range of settings, research and practice fields, as well as topics.

Also in 2017, the Global Working Group founded the **Society for Theory and Research on Salutogenesis (STARS)** (www.stars-society.org). This society aims to broadly advance and promote the science of salutogenesis in diverse fields. These principles guide it [5]:

- Transdisciplinarity: STARS connects scholars from diverse disciplines, who share an interest in the science of salutogenesis.
- Open membership: STARS welcomes anyone with an interest in the science of salutogenesis. Having published salutogenesis articles is not a condition of membership. Students are especially welcome. Joining is easy, and it is free of charge.
- Sharing: STARS members are encouraged to announce their publications, news and events on the STARS website and to download free materials.

Through this inviting, open membership, STARS meanwhile has over 2700 members from 90 countries. They apply Salutogenesis to diverse fields, including health care, everyday life, education, life stages, inter-cultural development, migration and politics. In the membership database, members can find colleagues from a specific field or geographic region for networking. STARS offers numerous free resources including the sense of coherence (SOC) instruments in various languages, a blog, news, video-recorded webinars, the original books by Antonovsky, recent publications, training support and a regular newsletter. The recent addition is a Lexicon of Salutogenesis. It covers key concepts of Salutogenesis and important conceptual developments since Antonovsky's original work.

Through STARS, the Global Working Group also organizes regular **International Conferences on Salutogenesis** open to all interested. The last conference in 2024 focused on "Everyday life and crises as opportunities for salutogenic transformation." It aimed to advance the vision of coherence-rich, thriving societies that master societal challenges in a humanistic way. Such challenges include pandemics, racism, rising inequalities, eroding democracies and planetary health. The conference aimed to investigate the role of coherence and value systems in the development of new equilibria of individuals, groups, organizations and societies in the face of such crises.

All these activities are supported by the **Center of Salutogenesis** at the University of Zürich (https://stars-society.org/center/). In 2017, the center was initiated by its current chair, Georg Bauer, and formally launched by the President of the University of Zürich. It is financially supported by a philanthropic foundation. The Center belongs to the "Institute of Epidemiology, Biostatistics and Prevention" at the University's Faculty of Medicine. The Center of Salutogenesis has the purpose of "putting salutogenesis to work." This purpose is implemented through two strategies. First, the Center supports the advancement of the overall concept, its dissemination and its application in various fields. It does so by hosting the STARS society and by supporting the Global Working Group. Second, the Center conducts research on salutogenic working life and salutogenic organizations. Currently, this includes research on Work-related SoC, positive health at work, as well as applying salutogenesis to the digital transformation of work, co-creation of health care and crafting job- and off-job life. The application of Salutogenesis to the specific context of working life facilitates testing and advancing this concept in general.

References

1. Lindström, B. (2022). Mileposts in the development of Salutogenesis. In M. Mittelmark, G. F. Bauer, L. Vaandrager, J. Pelikan, S. Sagy, M. Eriksson, B. Lindström, & C. Magistretti (Eds.), *The handbook of Salutogenesis* (2nd ed., pp. 5–9). Springer.
2. Lindström, B., & Eriksson, M. (2010). *The hitchhiker's guide to Salutogenesis. Salutogenic pathways to health promotion*. Folkhälsan Research Center, Health Promotion Research and the IUHPE Global Working Group on Salutogenesis.
3. Mittelmark, M. B., Sagy, S., Eriksson, M., Bauer, G. F., Pelikan, J. M., Lindström, B., et al. (Eds.). (2017). *The handbook of Salutogenesis*. Springer.
4. Mittelmark, M. B., Bauer, G. F., Vaandrager, L., Pelikan, J. M., Sagy, S., Eriksson, M., et al. (2022). *The handbook of Salutogenesis* (2nd ed.). Springer.
5. Bauer, G. F. (2022). Salutogenesis meeting places: The global working group, the center and the society on salutogenesis. In M. B. Mittelmark, G. F. Bauer, L. Vaandrager, J. Pelikan, S. Sagy, M. Eriksson, et al. (Eds.), *The handbook of Salutogenesis* (2nd ed., pp. 47–50). Springer.

Index

A
Activism, 103
Advances, 90, 130, 134, 135
Antonovsky, Aaron, 3, 4, 13–18, 21–28, 30, 35, 36, 42, 47–49, 51, 53, 80–82, 86, 89, 91–93, 95–97, 103, 106, 107, 109, 111, 117–120, 128, 129, 133, 134
Antonovsky, Avishai, 37, 122

C
Capacity building, 59, 122, 129, 130
Coherence, 76, 89–92, 94–97, 99, 109, 131, 135
Continuum, 15–17, 21, 27, 42, 49, 50, 80, 81, 85, 118, 120, 128, 129
Culture, 5, 9, 23, 25–26, 30, 39, 51, 93, 128

D
Definitions, 6, 7, 15, 24, 28, 42, 48–51, 75, 79, 97, 98, 104–106, 128, 129
Dialogues, 23, 28, 64–65, 73, 86, 91, 94, 98–100, 103, 108–111, 130
Dimensionality, 128

F
Flourishing/languishing, 49, 50, 119
Future, 4, 5, 23–25, 61, 96, 98–100, 118, 127–131

G
Generalized resistance resources (GRRs), 9, 21–27, 39, 49, 53, 60, 71, 72, 74, 75, 80, 92, 94, 107–110, 120, 129
Global development, 103, 104, 107–109, 111

H
Health, 3, 13, 21, 36, 47, 59, 70, 79, 90, 104, 118, 127
Health behaviours, 9, 53
Health promotion, 3–10, 17, 18, 22, 24–30, 47, 48, 50, 53, 59, 64–65, 72, 73, 79, 81, 82, 85, 92, 94, 97, 98, 103–111, 119–121, 123, 128–131, 133, 134
Healthy learning, 69–76
Healthy learning environments, 69–76
Human rights, 4, 5, 9, 10, 25, 48, 60, 64, 103, 104, 107–108, 110, 130, 131

I
Implementations, 6, 61–64, 106, 108, 111, 127
Interventions, 9, 22, 59–65, 85, 86, 119, 121, 122, 130

J
Job crafting, 84
Job demands, 80–83, 86
Job resources, 80–86

L

Learning processes, 5, 9, 18, 28, 30, 60, 64, 69–76, 89, 104, 110, 119, 123, 130

M

Measurement, 26, 27, 41, 128, 129
Mental health, 9, 22, 28, 47–53, 60, 61, 79, 85, 104, 106, 127, 129, 130
Mortality, 51

O

Orientation to life questionnaire, 35–42, 131
Ottawa Charter, 4–10, 25, 29, 47, 48, 64, 79, 83, 103–111, 119, 130, 134

P

Paradigm, 6, 17, 18, 106, 121, 122
Participation, 22, 29, 60, 61, 86, 92–94, 96, 98, 107–111, 120, 121, 130
Planetary health, 103–111, 129, 130, 135
Positive health, 6, 27, 81, 82, 128, 129, 135
Power dynamics, 120–121

Q

Quality of life (QoL), 4–9, 17, 18, 22, 25, 26, 28, 47–53, 61, 70, 76, 105, 106, 130

R

River of health and life, 17–18

S

SAL-questionnaires, 40
Salutogenesis, 3–10, 13, 15, 17, 18, 21, 22, 24–29, 36–38, 48, 49, 59–65, 69–76, 79–87, 89–100, 103, 104, 106, 107, 110, 127–131, 133–135
Salutogenic principles, 22, 60
Salutogenic questionnaires, 22, 40
Salutogenic theory, 21, 22, 24, 26, 27, 30, 35, 39, 42, 47, 49, 50, 59, 61, 64, 70, 97, 99, 104, 118–120, 122–124
Sense of coherence (SOC), 3, 6, 17, 21–23, 25–30, 35–42, 48, 51–53, 59–61, 70, 72, 75, 76, 80–86, 89, 91–96, 117–123, 128, 134
Settings approaches, 98
Social fragmentation, 100
Society, 6, 18, 21, 22, 26, 27, 37, 47, 48, 50, 52, 79, 85, 89–100, 119–122, 129–131, 133–135
Specific resistance resources (SRRs), 9, 21–26, 37, 39, 42, 49, 60, 64, 74, 110, 129
Stability, 25, 39–41, 119

W

Work, 18, 23, 25, 29, 37, 41, 48, 49, 52, 64, 70, 72, 74, 79–87, 93, 96, 98, 99, 118, 119, 130, 134, 135
Work-related Sense of Coherence (Work-SoC), 37, 83–85

If you have any concerns about our products,
you can contact us at:
productsafety@springernature.com

In case Publisher is established outside the EU,
the EU authorised representative is:
Springer Nature Customer Service Center GmbH
Europaplatz 3, 69115 Heidelberg, Germany

Printed by Elcograf S.p.A.
in Pomezia, Italy

If you have any concerns about our products,
you can contact us on
ProductSafety@springernature.com

In case Publisher is established outside the EU,
the EU authorized representative is:
**Springer Nature Customer Service Center GmbH
Europaplatz 3, 69115 Heidelberg, Germany**

Printed by Libri Plureos GmbH
in Hamburg, Germany